ACS SYMPOSIUM SERIES **816**

Bioactive Compounds in Foods

Effects of Processing and Storage

Tung-Ching Lee, Editor

Rutgers, The State University of New Jersey

Chi-Tang Ho, Editor

Rutgers, The State University of New Jersey

American Chemical Society, Washington, DC

Chemistry Library

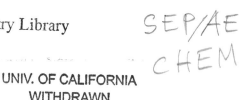

SEP/AE
CHEM

UNIV. OF CALIFORNIA
WITHDRAWN

Library of Congress Cataloging-in-Publication Data

Bioactive compounds in Foods : effects of processing and storage / Tung-Ching Lee, Chi-Tang Ho, editors.

p. cm.—(ACS symposium series ; 816)

Includes bibliographical references and index.

ISBN 0–8412–3765–4

1. Food analysis. 2. Bioactive compounds–Effect of temperature on. 3. Food industry and trade.

I. Lee, Tung-Ching, 1941 II. Ho, Chi-Tang, 1944- III. Series.

TP372.5 .B56   2002
664—dc21   2002022727

The paper used in this publication meets the minimum requirements of American National Standard for Information Sciences—Permanence of Paper for Printed Library Materials, ANSI Z39.48–1984.

PRINTED IN THE UNITED STATES OF AMERICA

ac*

Foreword

The ACS Symposium Series was first published in 1974 to provide a mechanism for publishing symposia quickly in book form. The purpose of the series is to publish timely, comprehensive books developed from ACS sponsored symposia based on current scientific research. Occasion-ally, books are developed from symposia sponsored by other organizations when the topic is of keen interest to the chemistry audience.

Before agreeing to publish a book, the proposed table of contents is reviewed for appropriate and comprehensive coverage and for interest to the audience. Some papers may be excluded to better focus the book; others may be added to provide comprehensiveness. When appropriate, overview or introductory chapters are added. Drafts of chapters are peer-reviewed prior to final acceptance or rejection, and manuscripts are prepared in camera-ready format.

As a rule, only original research papers and original review papers are included in the volumes. Verbatim reproductions of previously published papers are not accepted.

ACS Books Department

Contents

Preface

"We are what we eat" is truly relevant. If food processing is defined as including all treatment of foodstuffs from harvest to consumption, then more than 95% of our food may be considered as processed. The varied events and processes that take place during the preparation, processing, and storage of food is complicated due to the complex chemical heterogeneity of foods and the correspondingly complex reactions and processes that take place.

In general, thermal processing of foods is extremely beneficial, resulting in increases in digestibility, destruction of antagonists of vitamins and enzymes, and in many instances destruction of toxins that occur normally in foods. Food processing also leads to products that affect flavor, aroma, taste, nutrition, toxicologically and physiologically active compounds, and overall quality. In most cases, food processing and storage cause some reduction in the nutritional value of foods. The effect on bioactive compounds is the subject of this book.

Food processing causes changes to bioactive compounds in foods. Many bioactive health-promoting substances are unstable during processing and storage. They undergo various chemical reactions such as oxidation, hydrolysis, thermal degradation, and Maillard reaction; the result being a change and reduction in their bioactivity. On the other hand, processing can also generate new bioactive compounds. Some of these compounds, derived from the processing of popular food items such as garlic, soybean, tea, and dairy products, have been found to be beneficial in the prevention and treatment of various ailments and diseases.

Nowadays, bioactive nutraceuticals or functional foods are considered to be any food or part of a food that provides medical or health benefits, including prevention and treatment of a disease, and are widely considered to be critical for human health. Overwhelming evidence from epidemiological studies indicate that diets rich in specific foods (e.g., fruit and vegetables) are associated with a lower risk of several degenerative diseases. These results created a new perspective concerning the potential of diet in preventing serious diseases in the future. However, the health-promoting capacity of bioactive compounds in these foods strictly depends on their processing history. This aspect has been generally neglected or scarcely considered in clinical and epidemiological studies. Processing is expected to affect content, activity, and bioavailability of bioactive compounds and cannot be overlooked.

With better understanding of the mechanisms and kinetics of specific reactions relating to the stability of bioactive compounds during processing and storage, we may seek the possibility of modifying processing and storage procedures to minimize the undesirable effects and simultaneously maximize desirable effects. Furthermore, recent advances in emerging novel technology in food processing (e.g., non-thermal) and storage (e.g., modified atmosphere) are becoming more sophisticated and diverse, resulting in improved food quality including an increase in the retention of bioactive compounds after processing.

This book originated from a symposium entitled "Effect of Processing and Storage on Bioactive Compounds" held March 26–30, 2000 in San Francisco, California and was sponsored by the American Chemical Society (ACS) Division of Agricultural and Food Chemistry. The goal of the symposium was to bring together international research leaders with diverse backgrounds from academia, industry, and government to present their latest basic findings on processing effect on bioactive compounds in foods with up-to-date experimental data, mechanisms, methodologies, practical application to food quality, and the nutritional and clinical impacts to health benefits.

This book contains much pertinent and new information about current interests on the effect of processing on a variety of bioactive compounds in various foods. This book is intended to provide information for food scientists, food chemists, nutritionists, dietitians, medicinal and pharmaceutical scientists, physicians, and health-conscious consumers. It is our hope that this book will inspire food scientists and others who are engaged in related areas of research to produce higher quality and healthier foods to benefit all consumers through optimizing and modifying existing conventional processing and storage techniques as well as adoption of novel processing technologies. In addition, today's consumers can better understand how to avoid excessive nutrient losses during food preparation.

We thank the chapter authors for their efforts and time. We also extend our sincere gratitude to the many scientists who reviewed the chapters found herein. We thank Stacy VanDerWall, Kelly Dennis, and Anne Wilson in acquisitions; Margaret Brown in editing and production of the ACS Books Department; and Karen Ratzan of Rutgers University for providing persistent support.

We dedicate this book to the memory of Mary Ho, wife of Professor Ho, a writer who dedicated much of her work to food for health.

Tung-Ching Lee
Chi-Tang Ho
Department of Food Science and the Center
    for Advanced Food Technology
Rutgers, The State University of New Jersey
New Brunswick, New Jersey 08901–8250

Chapter 1

Marine Lipids as Affected by Processing and Their Quality Preservation by Natural Antioxidants

Fereidoon Shahidi[1] and Se-Kwon Kim[2]

[1]Department of Biochemistry, Memorial University of Newfoundland, St. John's, Newfoundland A1B 3X9, Canada
[2]Department of Chemistry, Pukyong University, Pusan 608–737, Korea

Interest in marine oils and seafoods has been intensified because of beneficial health effects ascribed to their long-chain polyunsaturated fatty acid (LCPUFA) components. These relate to the prevention and/or treatment of cardiovascular disease, rheumatoid arthritis and other disorders has been realized and led to their use in a variety of products. The LC PUFA in marine oils include eicosapentaenoic acid (EPA), docosahexaenoic acid (DHA) and to a lesser extent docosapentaenoic acid (DPA). Oils rich in LCPUFA originate from algal sources as well as the liver of lean fish, body of fatty fish and blubber of marine mammals. Due to their high degree of unsaturation, marine lipids and seafoods are highly susceptible to oxidation during processing and storage. Thus, their stabilization with antioxidants and use of appropriate packaging conditions is necessary. In addition, LCPUFA concentrates produced via physical, chemical or enzymatic processing provide specialty products that are of interest as nutraceuticals and pharmaceuticals.

1

The popularity of seafoods and marine products has increased over the last several years, not only because of their varied flavors and sensory qualities, but mainly due to perceived health benefits ascribed to them. The bioactive compounds from marine resources include oils (from fish, seal blubber, shark liver and algae) with relatively large content of long chain omega-3 polyunsaturated fatty acids, shark cartilage, chitin and chitosan, enzymes, proteins hydrolysates and peptides, vitamins and seaweed, among others (*1*).

This overview concentrates on describing some of the main sources of marine lipids and provides information about characteristics of marine oils with respect to their compositional characteristics and food and nutraceutical applications. Quality preservation of marine oils with natural antioxidants and preparation of omega-3 concentrates are also described.

Long-chain Omega-3 Polyunsaturated Fatty Acids (LC ω3 PUFA)

Marine oils originate from the body of fatty fish, liver of lean fish and the blubber of marine mammals (*2,3*). The common long chain ω3-PUFA in these oils are eicosapentaenoic acid (EPA), docosapentaenoic acid (DPA) and docosahexaenoic acid (DHA). These fatty acids are derived from α-linolenic acid mainly via a series of chain elongation and desaturation steps as well as a final proximal chain shortening step (Figure 1). While DPA is present in low levels in fish oils relative to EPA and DHA, it is found at about 5% in seal blubber oil. Both EPA and DHA are synthesized mainly by uni- and multicellular phytoplankton and algae and are eventually transferred through the food web into the lipids of aquatic species. The high content of ω3 PUFA in marine organisms is suggested to be a consequence of cold temperature adaptation in which ω3 PUFA remain liquid and oppose any tendency to crystallize.

In humans, the beneficial effects of LC ω3 PUFA in health are ascribed to their role in modifying the synthesis of eicosanoids. The beneficial effects of these fatty acids on coronary heart disease (CHD) are exerted via the lowering of plasma triacylglycerols, increasing the level of HDL and lowering of cholesterol levels, reducing the likelihood of cardiac arythmias and thrombotic diseases as well as lowering of blood pressure in individuals with high blood pressure (*4*). Other effects of marine oils relate to the relief of arthritis, improvement of diabetic condition and enhancement of body immunity. However, individuals on blood thinners must exercise caution in taking high levels of marine oils as this may cause sudden bleeding and delayed clotting. Certain consumers may also experience nausea and diarrhea.

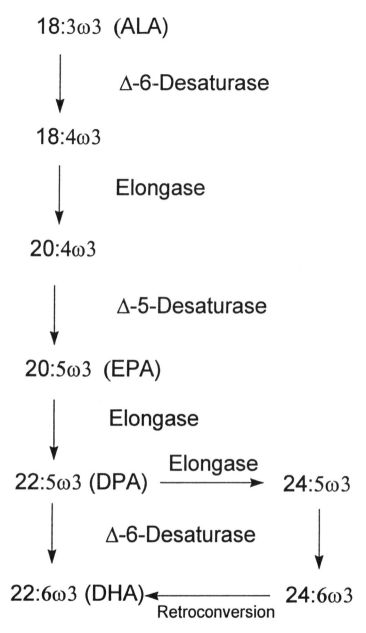

Figure 1. Formation and metabolism of essential fatty acids.

The very first observation on the beneficial effects of marine oils on CHD was noticed when the incidence of myocardinal infarction in Eskimos and Danes was compared (see Table I). While both group consumed about the same level of fat, there was a marked difference in the type of fatty acid constituents of lipids consumed. The ratio of ω3 to ω6 fatty acids in the Eskimos was nearly ten times higher than those of Danes, despite their higher level of cholesterol intake. There have also been several studies demonstrating the positive findings related to the long term intake of marine oils with respect to lowering of mortality from CHD (5).

Table I. Incidence of Myocardinal Infarction (MI) and Dietary Lipid Intake in Eskimos and Danes[1]

Parameter	Eskimos	Danes
MI	3	40
Energy from lipid (%)	39	42
ω6 - PUFA (g/day)	5	10
ω3 - PUFA (g/day)	14	3
ω3 / ω6	2.8	0.3
Cholesterol (mg)	790	420

[1]From ref. 5.

Another important beneficial effect of marine oils relates to their involvement in the development of brain and retina (6). These effects are specifically related to the DHA component of the oils. Fetal DHA status increases rapidly and exponentially during gestation, especially during the last trimester. Preterm infants may particularly suffer from lower levels of DHA and, if fed on formulas devoid of DHA, their DHA level would be adversely affected (7). However, breast feeding has been shown to improve the DHA status of infants, both preterm and full term. Among indicators evaluated were verbal, performance and IQ levels; all of which were significantly higher in breast-fed subject (8).

Table II shows the fatty acid composition of marine oils and an algal oil (DHASCO). The presence of varying proportions of EPA, DPA and DHA in these oils may be important in the intended application areas. Furthermore, the positional distribution of fatty acids in the triacylglycerols of such oils would undoubtedly affect their absorption and metabolism within the body.

Table II. Fatty Acid Profile of Seal Blubber, Cod Liver, Menhaden and Algal Oils

Fatty acid (w/w%)	Seal Blubber	Cod liver	Menhaden	Algal
14:0	3.73	3.33	8.32	14.9
14:1ω5	1.09	0.15	0.38	0.20
16:0	5.98	11.1	17.1	9.05
16:1ω7	18.0	7.85	11.4	2.20
17:0	0.92	0.61	2.45	-
17:1	0.55	0.44	1.86	-
18:0	0.88	3.89	3.33	0.20
18:1ω9	20.8	16.6	8.68	18.9
18:1ω11	5.22	4.56	3.46	-
18:2ω6	1.51	0.74	1.42	1.01
18:3ω6	0.19	0.22	0.51	-
18:3ω3	0.40	0.24	1.31	-
18:4ω3	1.00	0.61	2.90	-
20:0	0.11	0.05	0.20	-
20:1ω9	12.2	10.4	1.44	-
20:4ω6	0.46	0.22	0.83	-
20:5ω3	6.41	11.2	13.2	-
22:1ω11	2.01	9.07	0.12	-
22:2	0.37	-	0.02	-
22:5ω3	4.66	1.14	2.40	0.51
22:6ω3	7.58	14.8	10.1	47.4

In the particular case of seal blubber oil, it has a higher content of DHA as compared with EPA and contains nearly 5% DPA which is present only in small amounts in fish oils. Furthermore, the oil has most of its LC ω3 PUFA in the *sn*-1 and *sn*-3 positions as opposed to sn-2, and to a lesser extent, sn-3 in fish oils (2). A recent study, using rats, has shown superiority of seal blubber oil to fish oils in terms of its effect in reducing serum and liver triacylglycerols (9).

These authors further suggested that the hypotriacylglycerolemic effect of seal oil was caused by the suppression of fatty acid synthesis as the activities of fatty acid synthase, glucose-6-phosphate dehydrogenase and hepatic triacylglycerol lipase were significantly lower in rats fed seal oil than the group on fish oil (9).

Effect of Processing on Minor Constituents and Quality

During processing of edible oils, including marine oils, there is a significant loss of their minor constituents. These minor constituents are nontriacylglycerol components and represented by unsaponifiable matter (10). The unsaponifiable matter in oils generally includes a variety of active ingredients that could stabilize them against oxidative deterioration. The nontriacylglycerol constituents in marine oils belong primarily to the tocopherols/tocotrienols, sterols (including cholesterol), phospholipids, carotenoids, squalene, alcohols and waxes. Each marine oil may contain several classes of the above substances and the proportion of nontriacylglycerols in the oils could vary considerably. As an example, shark liver oil contains a relatively high amount of squalene.

Processing of marine oils generally includes rendering/extraction, refining, bleaching and deodorization (11). Each step may considerably reduce the content of minor compounds, similar to that observed for vegetable oils. Table III shows changes in the contents of α-tocopherol, free fatty acids (FFA) and squalene in seal blubber oil (SBO) during processing. The oils after alkali refining, bleaching and deodorization had 98.5% of their FFA removed as compared to those of the crude samples (from 1.37 to 0.04%). Industrial specifications indicate that the FFA content of processed marine oils should be less than 0.5% (12).

Tocopherol is another important minor component that may act as an antioxidant in marine oils. The tocopherol content of oils depends on the source material (organ, muscle, liver or blubber), PUFA content of the oil, as well as processing and storage conditions. The only tocopherol detected in seal blubber oil was α-tocopherol. According to Ackman and Cormier (13) and Kinsella (14), α-tocopherol is the major tocopherol present in marine oils. The content of α-tocopherol was very low and changes observed were mainly during the deodorization step. Tocopherols are heat labile and may also volatilize at high temperatures (100-200°C) reached during deodorization. Processing of the oil also affected its squalene content.

The oxidative state of the oils, as exemplified by seal blubber oil, was also affected by processing. The crude oil was found to be more stable over a period of 125h storage under Schaal oven condition at 65°C (Figure 2). Each 24h under Schaal oven condition in equivalent to one month of storage at room temperature. Removal of naturally-occurring antioxidants during refining, bleaching and deodorization is thought to be responsible for the observed trends (*15*).

Table III. Changes in Tocopherol, Acid Value and Squalene Content of Seal Blubber Oil during Processing

Sample	α-Tocopherol mg/100g	Acid Value mgKOH/g	Squalene %
Crude	2.8 ± 0.18	2.72 ± 0.10	0.59 ± 0.02
Alkali-refined	3.2 ± 0.11	0.08 ± 0.00	0.44 ± 0.03
Refined-bleached	3.1 ± 0.12	0.03 ± 0.01	0.36 ± 0.02
Refined-bleached-deodorized	2.4 ± 0.09	0.04 ± 0.01	0.28 ± 0.02

From ref. *15*.

Stabilization of Marine Oil with Natural Antioxidants

Marine oils are highly prone to oxidative deterioration because of their high content of polyunsaturated fatty acids with 5 or 6 double bonds. To stabilize marine oils, different methodologies might be employed. Oxidation generally requires oxygen, a catalyst and PUFAs that exist in the oil. Thus, effective elimination of oxygen as well as catalysts such as light and heavy metals is essential. In addition, antioxidants might be added to the oil in order to quench oxidation. However, hydrogenation of the oil, although effectively inhibiting oxidation, would negate the beneficial effects of LC ω3 PUFA as these components are lost during the process; a similar concern exists when using interestinfication with saturated lipids.

Amongst methods of choice for control of oxidation of marine oils are encapsulation, microencapsulation, packaging in dark containers, nitrogen purging as well as application of antioxidants or any combination of the above (*16*). However, experience has shown that encapsulation and microencapsulation are effective only for a certain period of time and this depends on storage condition of materials as dictated by temperature,

8

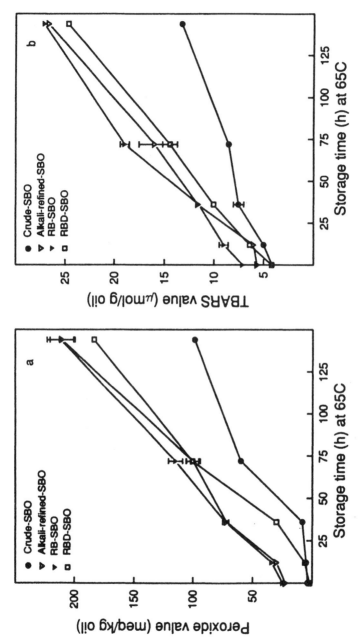

Figure 2. Stability of crude, alkali-refined, refined-bleached (RB) at refined-bleached-deodorized (RBD), seal blubber oil (SBO) as reflected in parasite values and 2-thiobarbituric acid reactive substances (TBARS) during storage.

moisture/humidity and presence of light. Our own experience has also indicated that encapsulation should be carried out under a nitrogen atmosphere and the type of gelatin with respect to its metal ion content be carefully considered.

The activity of natural antioxidants is related to their ability to scavenge free radicals, decompose free radicals, quench singlet oxygen and act as synergists with other antioxidants. Thus natural antioxidants with a number of active components, unlike single-component synthetic antioxidants, have multiple modes of action and may act as primary antioxidants, synergists, retarders, metal scavengers or singlet oxygen quenchers depending on the origin and nature of the material (17). Of the natural antioxidants, several preparations are now commercially available for use in foods. These include tocopherols, tocotrienols, ascorbic acid, its fat soluble counterpart ascorbyl palmitate and its isomer erythorbic acid and their salts, spices, spice extracts and oleoresins such as rosemary and sage and green tea extracts. However, presence of photosensitizers such as chorophyles in the extracts might compromise their efficacy as light-induced oxidation would lead to the formation of oxidation products. Therefore, elimination of light by encapsulation and storage in dark bottles may be considered (18).

Amongst antioxidants that have been used in stabilization of marine oils, synthetic antioxidants, mixed tocopherols and certain natural extracts have shown promise (19,20). However, presence of photosensitizers such as chlorophyll in the extracts might compromise their efficacy as light-induced oxidation would lead to the formation of oxidation of products, therefore elimination of light by encapsulation, use of dark bottles or complete dechlorophyllization of preparations may be considered (21). Table IV shows the effect of selected antioxidants on the stability of seal blubber and menhaden oils as reflected in the inhibition of formation of thiobarbituric acid reactive substances.

Food and Nutraceutical Applications of Marine Oils

Marine oils may be used in food formulations in order to provide the body with its ω3 PUFA requirement. In most cases, it is of interest to use the oil in products that are usually consumed within a few days from the time of preparation or the time of opening the container/package. Thus, use of such oils in bread, crackers, cereals and cereal bars, milk, fruit juices, salad dressing, mayonnaise and pasta as well as infant formula might be most desirable (Table V) (22,23). In these cases, depending on the time interval between preparation and consumption, application of antioxidants might prove beneficial. The choice of antioxidant would then depend on the nature of the system as it relates to bulk or continuous oil system versus emulsion and in the case of cereals and dried foods, the relative moisture content of the products.

**Table IV. Percent Inhibition of Oxidation of Marine Oils by
Antioxidants under Accelerated Storage Condition in Open Containers**

Antioxidant (ppm)	Seal blubber oil	Menhaden oil
DGTE (200)	32.9	32.9
DGTE (1000)	47.1	44.6
EC (200)	39.5	41.1
EGC (200)	40.5	46.5
ECG (200)	58.6	52.9
EGCG (200)	50.0	49.4
α-Tocopherol (200)	14.2	14.1
BHA (200)	23.0	25.3
BHT (200)	35.5	31.4
TBHQ (200)	56.3	51.8

Abbreviations are: DGTE, dechlorophyllized green tea extract; EC, epicatechin;
EGC, epigallocatechin; ECG, epicatechin gallate; and EGCG, epigallocatechin
gallate. BHA, butylated hydroxyanisole; BHT, butylated hydroxytoluene; TBHQ,
tert-butyl hydroquinone.

Table V. Food Application of Marine Oils

Cereals	Milk
Bread/hard bread	Orange juice
Crackers	Mayonnaise
Bars	Salad dressing
Pasta	Margarine
Infant formula	Spreads

As nutraceuticals, omega-3 concentrates are often used. A summary of the methodologies involved in production of omega-3 concentrates is provided below.

Omega-3 Concentrates and Their Production

To prepare omega-3 concentrates, marine oils may be used as the starting material (23). Depending on specific needs and intended use of products, concentrates may be prepared in the form of free fatty acid, simple alkyl ester as well as acylglycerol. In addition, both physical and chemical procedures may be employed for producing omega-3 concentrates. These include, fractional crystallization, supercritical extraction, urea complexation, enzymatic hydrolysis, chromatographic procedures, among others. Furthermore, products may be prepared in such a way that they contain different proportions of EPA, DPA and DHA. Innovative preparation of such products is most important. In general, for heart patients, concentrates containing high amounts of EPA is prescribed while for infant formula application, DHA is most beneficial.

Fatty Acid Concentrates

Fatty acids may be produced via the hydrolysis of the oil. Saturated and monounsaturated lipids may then be eliminated using distillation, urea complexation (24), low temperature crystallization (25) or chromatography (26) as well as supercritical fluid extraction (27). Due to the existing physico-chemical differences between highly unsaturated fatty acids (HUFA) and other fatty acids present, separation could be achieved, however, the recovery of the concentrates is dictated by the methodology employed as well as the availability of the equipment and consideration of the economy of the process. Furthermore, concentrates may be prepared using a multistage process using one or more techniques. Generally, lower yields are obtained as concentrates with higher contents of HUFA are prepared.

Alkyl Ester Concentrates

Simple alkyl esters of HUFA may be prepared by esterification of the free fatty acids with methanol or ethanol. Most commercial concentrates were traditionally in the form of alkyl esters.

Acylglycerol Concentrates

Acylglycerol concentrates of omega-3 fatty acids may be prepared using chemical or enzymatic procedures. It is possible to use enzymes to selectively hydrolyse saturated and monounsaturated lipids from the triacylglycerols, hence preparing a concentrate. Concentrates prepared in this manner may contain varying levels of omega-3 fatty acids depending on the source material, the enzyme used and process variables. For seal blubber oil concentrates with approximately 40% omega-3 fatty acids were prepared (*28-30*).

Another method for production of concentrates in the acylglycerol form is by esterification of the omega-3 fatty acid or alkyl ester concentrates with glycerol (*31*). Both chemical and enzymatic procedures may be employed. The products may contain some monoacyl- and diacylglycerols, but triacylglycerols are usually dominant. Unreacted fatty acids or their alkyl esters may be recovered.

References

1. Ohshima, T. *Food Technol.* **1998**, *52*(6), 50-54.

2. Shahidi, F. In *Seal Fishery and Product Development.* ScienceTech Publishing Co.: St. John's, NF, 1998, pp. 99-146.

3. Shahidi, F. In *Functional Foods: Biochemicals and Processing Aspects.* Ed. G. Mazza. Technomic Publishing Co.: Lancaster, PA, 1998, pp. 381-401.

4. Osterud, B.; Elvevoll, E.; Barstad, H.; Brox, J.; Halvorsen, H.; Lia, K.; Olson, J.O.; Olson, R.L.; Sissener, C.; Rekdal, O.; Vegnild, E. *Lipids* **1995**, *39*, 1111-1118.

5. Dyerberg, J.; Bang, H.D.; Hgorne, H. *Acta Med. Scand.* **1976**, *200*, 69-73.

6. Conner, W.E.; Neuringer, M.; Reisbick, S. *Nutr. Res.* **1992**, *50*(4), 21-29.

7. Hornstra, G.; Al, M.D.M.; van Houwelington, A.C.; Foreman-van Drongelen, M.M. *Eur. J. Obster. Gynecol. Reprod. Biol.* **1995**, *61*, 57-62.

8. Lucas, A.; Morley, R.; Cole, T.J.; Lister, G.; Leeson-Payne, C. *Lancet* **1992**, *339*, 261-264.

9. Yoshida, H.; Mawatari, M.; Ikeda, I.; Imaizumi, K.; Seto, A.; Tsuji, H. *J. Nutr. Sci. Vitaminol.* **1999**, *45*, 411-421.

10. Shahidi, F.; Shukla, V.K.S. *INFORM* **1996**, *7*, 1227-1232.

11. Shahidi, F.; Wanasundara, P.K.J.P.D.; Wanasundara, U.N. *J. Food Lipids* **1997**, *4*, 199-231.

12. Bimbo, A.P. *J. Am. Oil Chem. Soc.* **1987**, *64*, 706-715.

13. Ackman, R.G.; Cormier, A. *J. Fish Res. Bd. Canada* **1967**, *24*, 357-373.

14. Kinsella, J.E. *Seafoods and Fish Oils in Human Health and Disease.* Marcel Dekker, New York, 1987, pp. 231-237.

15. Shahidi, F.; Wanasundara, U.N. *J. Food Lipids* **1999**, *6*, 159-172.

16. Wanasundara, U.N.; Shahidi, F. *Food Chem.* **1998**, *63*, 335-342.

17. Boyd, L.C. 1999. 217th National Meeting of the American Chemical Society, Anaheim, CA, March 21-25, Abstract AGFD # 11.

18. Shahidi, F.; Wanasundara, U.N. In *Flavor Technology, Physical Chemistry, Modification and Process.* Ho, C-T.; Tan, C-T.; Tong, C-H., Eds.; ACS Symposium Series 610, American Chemical Society: Washington, D.C. 1995, pp. 139-151.

19. Shahidi, F.; Wanasundara, U.N.; He, Y.; Shukla, V.K.S. In *Flavor and Lipid Chemistry of Seafoods.* ACS Symposium Series 674. Shahidi, F.; Cadwallader, K.R., Eds.; American Chemical Society: Washington, DC. 1997, pp. 186-197.

20. Wanasundara, U.N.; Shahidi, F. *J. Food Lipids* **1998**, *5*, 183-196.

21. Wanasundara, U.N.; Shahidi, F. *J. Food Lipids* **1995**, *2*, 73-86.

22. Bimbo, A.P. In *Marine Biogenic Lipids, Fats and Oils.* Ed. R.G. Ackman, CRC Press Inc.: Boca Raton, FL. 1989, pp. 401-433.

23. Newton, I.S. *Food Technol.* **1996**, *7*(2), 217-236.

24. Wanasundara, U.N.; Shahidi, F. *Food Chem.* **1999**, *65*, 41-49.

25. Stout, V.F.; Nilsson, W.B.; Krzynowek, J.; Schlenk, H. In *Fish Oil in Nutrition.* Stansby, M.E. Ed., Van Nostrand Reinhold: New York, NY. 1990, pp. 73-119.

26. Teshima, S.; Kanazawa, A.; Tokiwa, S. *Bull. J. Soc. Sci. Fish* **1978**, *44*, 927-930.

27. Mishra, V.K.; Temelli, F.; Ooraikul, B. *Food Res. Inst.* **1993**, *26*, 217-226.

28. Wanasundara, U.N.; Shahidi, F. *J. Am. Oil Chem. Soc.* **1998**, *75*, 1767-1774.

29. Wanasundara, U.N.; Shahidi, F. *J. Am. Oil Chem. Soc.* **1998**, *75*, 945-951.

30. He, Y.; Shahidi, F. *J. Am. Oil Chem. Soc.* **1997**, *74*, 1133-1136.

31. Shahidi, F.; Wanasundara, U.N. *Trends Food Sci. Technol.* **1998**, *9*, 230-240.

Chapter 2

Effect of Non-Thermal Treatments and Storage on Bioactive Compounds

Dietrich Knorr

Department of Food Biotechnology and Food Process Engineering, Berlin University of Technology, Königin Luise Strasse 22, D–14195 Berlin, Germany,

High hydrostatic pressure and high intensity electric field pulses are effective tools to process and modify foods and food constituents. Within the context of the current research and development priorities towards the development of health related, functional foods the ability of these technologies to more effectively recover or retain desirable bioactive compounds is especially attractive. Optimum use of the advantages and benefits of these technologies can also lead to new process designs and can contribute to the effective improvement and reorientation of existing process technologies.

Among the ambient temperature processes used for the preservation and modification of foods fermentation technologies, membrane processes, high pressure assisted processes, high intensity electric field pulse treatments and irradiation technologies are the most relevant ones. Fermentation results in biotransformations which can be a limitation when retaining the original properties of the raw materials is desired. The utilization of membrane processes is limited to liquid systems. Irradiation is applicable only to a restricted number of food materials and its widespread distribution is hindered by consumer concerns and opposition. On the other hand, high pressure treatment and the application of high intensity electric field pulses are applicable to solid as well as to liquid systems and can be utilized for preservation and modification of foods as well as for re-designing existing food processes or creating new ones.

High Hydrostatic Pressure Processing

For the application of high hydrostatic pressure the principle of Le Chatelier applies, according to which, under equilibrium conditions, a process associated with a decrease in volume is favored by pressure and vice versa. This is applicable to solid and liquid systems. Pressure primarily affects the volume of a system. The influence of pressure on the reaction rate can be described by the transition-state theory, where the rate constant of a reaction in a liquid phase is proportional to the quasi equilibrium constant for the formation of active reactants. Based on this assumption, at isothermal conditions, the pressure dependence of the reaction velocity constant is due to the activation volume of the reaction (1). The temperature dependent compressibility of water, one of the most important food constituents, ranges from 4% at 100 MPa and 22°C to 15% at 600 MPa and 22°C.

Adiabatic compression of water increases the temperature by approx. 3 °C per 100 MPa. Self ionization of water is also promoted by high pressure thus lowering the pH. Phase transition of water can be performed under pressure. At approx.1,000 MPa water freezes at room temperature, whereas the freezing point can be lowered to -22 °C at 207.5 MPa (1). As evident from the phase diagram of water and proteins under pressure given in Figure 1, storage of food under subzero temperatures without ice crystal formation, pressure shift freezing with instant and small ice crystal formation, supercooling or freezing along the phase transition line, or fast thawing of foods by pressurization is possible and offers attractive advantages over conventional processes such as instant ice crystal formation throughout the product (2). In addition, the phase diagram of proteins indicates the unique relationship between pressure and temperature regarding the transition from native to denatured state (Figure 1).

Most of the food related high pressure research has concentrated on the use of high pressure for preservation purposes especially on microbial inactivation (3). Current evidence suggests that vegetative microorganisms can be inactivated sufficiently effective (4) although systematic studies on inactivation kinetics, on pathogens and on inactivation mechanisms involved, including recovery phenomena of injured cells, are still scarce. It is encouraging to note that it was made possible to inactivate bacterial spores by taking advantage of the additional adiabatic heating of a combined pressure - temperature process while retaining food quality and functionality (5-7).

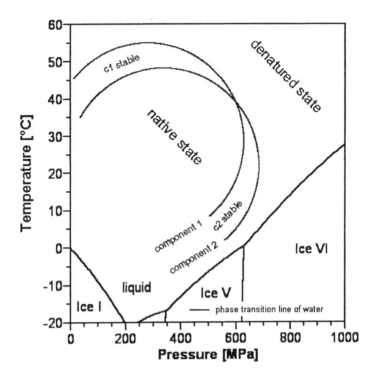

Figure 1. Phase diagram of water and proteins under pressure

Pressure Effects on Enzymes

Enzyme activities play an essential part in in food preservation processes and their control and monitoring is an integral part of process design and product development involving non-thermal technologies. The specificity of enzymatic reactions may be affected by pressure in various ways: (i) Pressurization at room temperature may bring about reversible or irreversible, partial or complete enzyme inactivation resulting from conformational changes in the protein structure and depending on the type of enzyme, micro-environmental conditions, pressure, temperature and processing time. (ii) Enzymatic reactions may be enhanced or inhibited by pressure, depending on the volume change associated with the reaction. Pressure-induced changes in the catalytic rate may be due to changes in the enzyme-substrate interaction, changes in the reaction mechanisms or the effect of a particular rate-limiting step on the overall catalytic rate. (iii) A macromolecular substrate (e.g. starch) may become more sensitive to enzymatic action once it has been unfolded or gelatinized by pressurization. (iv) Provided the cell membranes are altered, intracellular enzymes may be induced, activated and released hereby facilitating enzyme-substrate interactions (8).

Because of the eminent role of polyphenoloxidase (PPO) in food preservation, quality retention and quality enhancement and because of the extensive work being carried out with PPO, it is used as a representative for food related enzymes. Some of the data presented in the literature are summarized in Table 1.

Table 1. Effect of pressure on PPO from different origins (8,9)

As evident from the data given in Table 1 the information available is mostly fragmentary and only few kinetic data have been compiled. So far it can be concluded that PPO may either display inactivation or enhanced catalytic activity. The pressure level required for PPO inactivation is strongly dependent on origin and the pH of the medium. Also other micro-environmental conditions such as the presence of salts, sugars, and other additives have proven to influence the pressure stability of PPO (9). The complexity of the impact of pressure on a cellular level is exemplified in Figure 2.

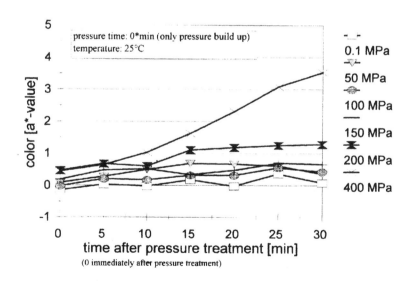

Figure 2A. Impact of pressure on time dependent browning development in potato cell cultures (10).

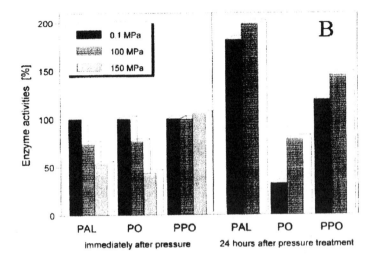

Figure 2B. Impact of pressure on enzyme activities in potato cell cultures(10).

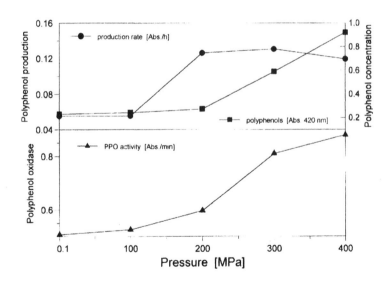

Figure 2C. Impact of pressure on polyphenol concentration, polyphenol production rate and PPO activity in potato culture extracts.(10).

It should also be noted that moderate pressure treatment (100 to 150 MPa) of potato (*Solanum tuberosum*) cell cultures initiated a stress response which resulted in a time dependent (after pressure treatment) induction of activities of PPO, peroxidase (PO) and phenylalanine ammonia lyase (PAL)(11).

Pressure Effects on Macromolecules

Several studies have shown that pressure treatment of vegetables can cause both firming (12-14) and softening of texture (15,16). Currently it is not entirely clear what mechanisms are responsible for the textural changes. In intact cells, polygalacturonidase or pectin methyl esterase are bound to the cell walls. Upon high pressure treatment the enzymes are likely to be liberated or activated. Data in Figure 3 demonstrate such an effect.

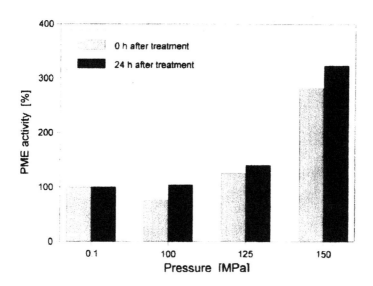

Figure 3. Effect of various high pressure levels on pectin methylesterase activity in tomato cultures (17)

It has been suggested that hardening of some plants by high pressure treatment is the result of insufficient inactivation of liberated pectin methylesterase (15). It is also believed that the pressure induced release of Ca^{++} from plant membranes plays an actvie role in the firming of certain plant tissues

Table 1. Effect of pressure on PPO from different origins (8,9)

Origin of PPO	Medium	Process conditions			Activation (A) Inactivation (I)
		MPa	°C	min	
Apple					
cv. NM	PB (pH 4.5, 0.5M)	150-300	25	1	A
		≥500	25	1	I
	PB (pH 5.4, 0.5M)	<300	25	1	A
		>700	25	1	I
	PB (pH 7.0, 0.5M)	≥100	25	1	I
Cv. Golden	crude extract (pH 7)	500	25	10	A
	crude extract (pH 3-6)	500	25	10	I
	cell free extract (pH 3.9)	>250	25	10	I
	puree	≥1000	20	5	I
	PB (pH 6.0, 0.1M)	≥600	25	5-60	I
cv. Granny Smith	cell free extract (pH 3.5)	≥125	25	10	I
cv. Jovagolden	cell free extract (pH 3.9)	≥250	25	10	I
cv. Red Chief	cell free extract (pH 4.1)	≥250	25	10	I
Apricot					
cv. NM	cell free extract (pH 3.9)	≥125	25	15	I
cv. NM	puree	300	20	2.5	A
	·	500	20	5	A
		700	20	2.5-10	I
		900	20	2.5-10	I
		300-500	50	2.5-10	I
		700-900	50	2.5-10	I
Avocado					
cv. NM	PB (pH 7.0, 0.1M)	≥800	25	>60	I
cv. NM	puree	350-700	20	10-30	I
Banana					
cv. NM	puree	520-690	20	10	no I
Carrot					
cv. NM	PB (pH 4.5-7)	100-800	20	1	A
		900	20	1	I
Grape					
cv. Chung shan	puree (pH 3.8)	≥600	25	15	I

Table 1. *Continued*

Onion					
cv. Stuttgarter Riese	TB (pH 6.5, 0.01M)	≤700	25	10	A
Pear					
cv. Barlett	slices	400	25	10	A
	cell free extract	200-600	25	10	A
cv. La France	PB (pH 7.0, 0.02M)	400-700	20	≥3	A
cv. Durondeau	PB (pH 7.0, 0.1M)	900	250	200	I
Plum					
cv. Mirabelle	cell free extract (pH 4.6)	≤500	25	15	no I
cv. Red Beauty	PB (pH 7.0, 0.1M)	900	25	180	no I
Potato					
cv. NM	cubes	400	25	25	no I
cv. NM	crude extract	≥200	25	10	I
cv. Desirée	cell culture	100-200	25	5-10	A
cv. Desirée	water surrounded cubes	400	20	15	I
Strawberry					
cv. NM	crude extract (pH 2-5)	500	25	10	I
	crude extract (pH 6)	500	25	10	A
	cell free extract (pH 3.8)	≥375	25	15	I
cv. Pajaro	puree (pH 3.9)	50-285	20	15	I
Tomato					
cv. Pera	puree (pH4.1)	<200	20	15	A
		200-500	25	15	I
Mushroom					
strain NM	PB (pH 6.5, 0.1M)	≥750	25	20	I
strain NM	PB (pH 6.5, 0.05M)	>100	25	>1	I
strain NM	crude extract	200	25	10	I
		400	25	10	A
		≥600	25	10	I
strain NM	NM	300-700	20	>10	A
strain NM	intact tissue	600	25	5	A
		800	25	5	I

22

(14). Interestingly, this hardening effects do not only occur immediately during and after pressure application but can also develop over extended storage of the treated products (Figure 4).

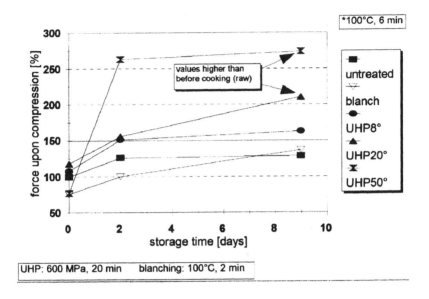

Figure 4. Texture changes of untreated, blanched and pressure treated (8°C, 20°C, 50°C) peas during storage at 4°C and subsequent cooking (18).

Recent data (19) show that pressure induced gels from certain fruit purees can withstand conventional sterilization regimes without loss in gel performance. Systematic studies are currently being carried out in our laboratory to better understand the interaction between food system composition, high pressure and resulting physico-chemical properties such as texture changes to possibly use high pressure for effective food structure engineering.

The potential of high pressure processing for animal derived products especially meat was extensively reviewed by Cheftel and Culioli (20) and will, therefore, not to be discussed further.

Pressure and Storage Effects on Micromolecules

An unique concept of combining high pressure treatment (150 MPa) and modified atmosphere packaging of fresh Atlantic salmon using 50% O_2 and

50% CO_2 did extend the microbial shelf life of the products for 4 days but promoted a detrimental effect on color change in the balance of oxidative rancidity (21). The changes in color values of liquid whole egg as related to pressure and temperature of treatment is demonstrated in Figure 5.

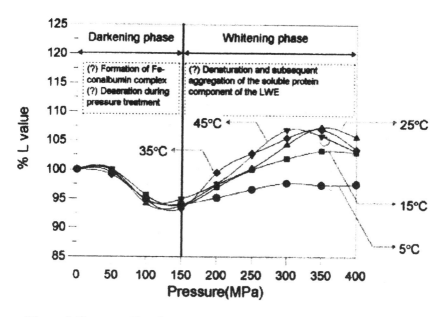

Figure 5.Changes of L-values of liquid whole egg with pressure at different temperatures (22.)

Sensory qualities of pressure treated orange juice were found to be closer to those of freshly squeezed one than to heat treated samples (23-25). This high quality of the pressure pasteurized products was retained unchangebly during storage for 1 to 3 month at 5°C. However, storage at 25°C resulted in rapid quality deterioration, due to dissolved oxygen and remaining enzyme activities (26,27). Similar effects were noticed during storage of strawberry jam for 30 days during which ascorbic acid concentration of the pressure treated samples was significantly reduced. The initially high concentration of dissolved oxygen (pressure treated:1.96 mg/100 g, heat treated: 0.39 mg/100 g) for the pressure treated samples was reduced to 0.17 mg/100 g during storage at 25°C for both treatments (27).

The effects of storage time and storage temperature on the ascorbic acid concentration of peas is presented in Figure 6. Pressure treatment of broccoli in CO_2 atmosphere proved to completely retain ascorbic acid during pressure treatment (Figure 7).

24

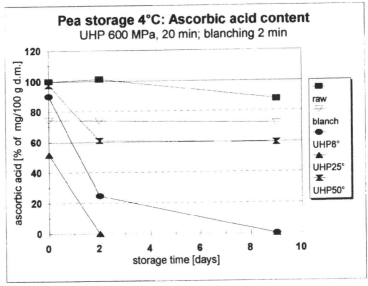

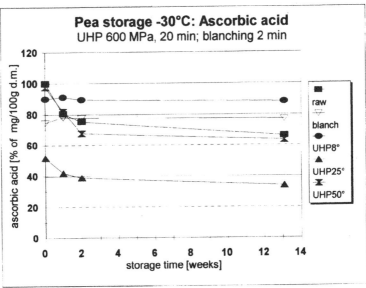

Figure 6. Effect of storage time and storage temperature on the ascorbic acid concentration of raw, blanched and pressure treated (8°C, 25°C, 50°C) peas (28)

It is also noteworthy that ascorbic acid losses of potato cubes due to leaching and thermal treatment during conventional blanching could be reduced significantly by high pressure blanching (29) which, in contrast to conventional thermal blanching, can be considered as waste free technology. Finally, sensory properties of onion and tomato juices were largely affected by pressure treatment, with tomato juice proving inedible owing to a strong rancid taste (30, 31). These three examples clearly demonstrate that successful pressure treatment requires a clear understanding of all factors involved and that effective removal of oxygen prior processing and inactivation of undesirable enzymes prior or during pressure processing is necessary. On the other hand, controlled monitoring of pressure supported stress reaction of plants can lead to increased productivity of the "plant bioreactor systems" for the enhanced biosynthesis of desired plant metabolites (32).

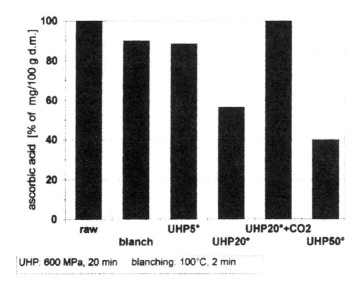

Figure 7. Ascorbic acid concentration of raw, blanched and high pressure treated broccoli (5°, 20°, 50°C) packaged in air or in CO_2 (20°C) (28).

Future Research Needs

From the few and selected examples summarized above it becomes evident that more kinetic data are needed on the impact of pressure on bioactive components in various food systems. For example the role of water in foods has not been studied systematically and a clear understanding of enzyme (in)activation is required. For example, increased enzyme activities during the pressure build up phase resulting in bioconversions in the products are being observed, although no enzyme activity can be obtained after pressure treatments.

Since most of the previous high pressure work concentrated on preservation effects, careful studies regarding nutrients, antinutrients, nutritional stress factors, allergens and toxis are necessary in future research including reaction kinetics during pressure treatment and during subsequent storage. Due to the complexity of food systems work on a cellular level using model systems such as cultured plant or animal cells will be a prime prerequisit for a better understanding of the mechanisms involved regarding the impact of high pressure on foods and food constituents.

High intensity electric field pulses

Applying an external electrical field to biological materials can induce a critical electrical potential across the cell membranes, which leads to rapid electrical breakdown and local structural changes of the cell membrane. This first field effect results in a drastic increase in permeability due to the appearance of pores in the membranes (33). This phenomenon has spurred work on many applications in the biosciences for the reversible or irreversible permeabilization of various biological membranes (34).). Most of the work regarding food systems has concetrated on the inactivation of microorganisms (35). From the data available so far it can be concluded that most vegetative microbial cells can be sufficiently inactivated by electric field pulses but no significant impact of electric field pulses on spores could be detected with the exception of induction of germination of *Bacillus subtilis* spores when treated in media of pH 5.0 and lower (36).

The irreversible permeabilization of cell membranes in biological materials offers a wide range of process application where cell membrane disruption is required, including expression, extraction or diffusion of plant metabolites as well as affecting heat and mass transfer of the products (37). Evaluation of the time sequence of electro-permeabilization at the cell membrane level led to the identification of membrane breakdown points for various food systems and also revealed that resealing of membranes takes place within seconds after the termination of the pulse (33) with pulse widths generally in the range of microseconds to milliseconds. This reversible membrane permeabilization termed electroporation in the field of plant genetics and applied for the insertion of genetic materials into plant cells could possibly be used for supporting the infusion of desirable biochemicals into food tissues. Additionally, work is been carried out in the authors laboratory to also use electrical field pulses at subcritical levels as stressor for plant and animal tissues and for microbial cells in order to induce enhanced production of desirable metabolites.

Energy requirements for membrane permeabilization of plant tissues by electric field pulses are in the range of 6 to 10 kJ/kg material as compared to approx. 250 to 300 kJ/kg necessary for thermal permeabilization of membranes (37). This is of importance because heat labile food constituents can be retained due to the possibility of low temperature application and because of the short process duration of high intensity electric field pulse treatments.

Stress Responses to Electric Field Pulses

Controlled sublethal stresses subjected to plant cells via pulsed electric fields have been applied to induce the enhanced production of desirable plant metabolites, especially antioxidants and antimicrobials. An airlift reactor equipped with coaxially placed electrodes was designed and is currently being used to identify optimum treatment and culture conditions (38). Post-treatment effects after subcritical, critical and supercritical membrane permeabilization have been monitored and are presented in Figure 8.

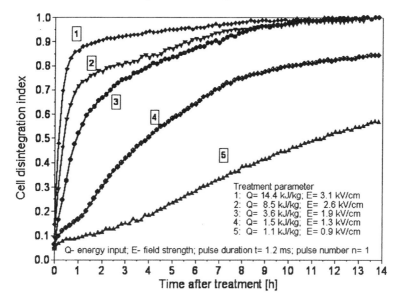

Figure 8. Effects of various total energy inputs via pulsed electric fields on the time dependent permeabilization kinetics of potato cells (Zp 1.0=100% permeabilization) (37)

Interestingly, electric field pulse treatment also induced a time-dependent increase of the degree of permeabilization. This is most likely due to induction of enzyme activities which also result in degradation of plant cell wall biopolymers such as pectins. Such degradation products due to responses to stressors like high pressure or electric field pulses seem to induce time-delayed biosynthetic activities from the plant cells resulting, for example, in the production of antimicrobials (39).

Application of Irreversible Membrane Permeabilization by Electric Field Pulse Treatment

Data on sensory analyses of various food products pasteurized either by heat or by electric fields indicated that electric field pulse treated samples (e.g. apple juice, orange juice, milk, whole liquid egg, pea soup) resulted in foods with sensory qualities being favorable over heat processed ones (35).

Increases in juice yields of carrots, apples, grapes and black currants after electric field pulse treatment as compared to untreated juices and yields comparable to those treated with commercial pectolytic enzyme preparations have been reported (40-43). The electric field pulse treated products resembled the composition of the freshly pressed juices closer than the enzyme treated ones due to the short processing time and due to the physical permeabilization of membranes which does not involve the degradation of biopolymers as experienced in enzyme treated juices. It should also be noted that by varying the process conditions during electric field pulse treatment of the raw materials the composition (e.g. acidity, turbidity, minerals, pectin) of the resulting juices could be influenced further (41-43). A simplified summary of the impact of electric field pulse treatment on juice compositions is provided in Table 2.

In light of the current interest in fortified fruit juices and considering the importance of minerals in functional foods (44), these findings can provide the basis for the development of new process improved functional food products. Ade-Omowaye et al. (45) reported significantly higher milk yields after field pulse treatment of coconuts as compared to mechanically disrupted controls. The ß-carotene concentration of juice resulting from electric field pulse treated red paprika was approx. 1/3 higher than that of the untreated and of the enzyme pre-treated samples. Juice yields of enzyme pre-treated and electric field pulse treated samples were similar, but significantly (12%) higher than the untreated control (46).

High intensity electric field pulses have also been used effectively and successfully for the pre-treatment of food systems prior to extraction and dehydration (47). Applying electric field pulse treatment to sugar beet processing allows to reduce the extraction temperatures from approx.80°C to

45°C and makes it possible to eliminate extraction completely and to recover sugar beet juice by pressing only without loss in sucrose yield (48).

Table 2. Effects of high electric field pulse treatment on the composition of various fruit juices as compared to untreated ones (41-43)

	Grapes	Apple	Black-currant
Total solids (°B)	↑	↑	→
Dencity (g/ml)	↑	→	→
Acidity (mval/l)	↑	↑	↑
pH-value	↓	↓	→
Conductivity (mS/cm)	↑	↑	↑
Turbidity	↓	↑	↓
Saccharose		↑	↑
Glucose	↑	→	→
Fructose	↑	→	↓
Minerals (Ca, K)	↑	↑	↑
Pectin	↓	↑	↑
Protein	↑	→	↑
Ascorbic acid			↑
Tartaric acid	↑		

Future Research Needs

High intensity electric field pulses have a great potential for process and product development beyond their use for pasteurization purposes. As indicated above the possibilities for inducing stress reactions, for monitoring reversible permeabilization as well as the ability for energy efficient, short time processing leading to irreversible permeabilzation of biological membranes allow numerous future applications. However, to be able to effectively and successfully utilize such a potential, a better understanding of the impact of field pulses on the cellular level (including effect on cell walls and cell wall constituents), on the interaction between electric field pulses and complex food systems, as well as the compilation of kinetic data on changes occuring in food systems during and after pulsed electric field treatment is required. Further, treatment chamber designs and electric field distributions need to be optimized, interactions between electrode materials and foods must be examined and systematic work regarding safety issues (e.g. on free radical formation) needs to be carried out to assure regulatory approval.

Acknowledgements

Part of this work have been carried out with support from the European Commission (Grant No. FAIR CT97-3044), the German Academic Exchange Service (DAAD), the German Industrial Research Foundation (AIF No.12516NI) and the German Ministery for Education and Research (BMBF No.0339928).

References

1. Tauscher,B. Pasteurization of food by high hydrostatic pressure:chemical apects.Z.Lebensm.Unters.Forsch.1995,200,3-13.
2. Denys,S; Schlüter,O;Hendrickx,M;Knorr,D. Impact of high hydrostatic pressure on water-ice transitions in foos.in Hendrickx,M;Knorr,D.High Pressure Treatment of Foods, Aspen Publishers Inc., Gaithersburg,MD (in press).
3. Palou,E; Lopez-Malo,A; Barbosa-Canovas,G.V;Swanson,B.G. High-pressure treatment in food preservation.in. Rahman,M.S. Handbook of Food Preservation. Marcel Dekker, Inc. New York,N.Y;p.533.
4. FDA. Kinetics of Microbial Inactivation for Alternative Food Processing Technologies. Center for Food Safety and Applied Nutrition, Summit-Argo,Ill.
5. Meyer,R. Ultra high pressure, high temperature food presevation process. US patent 6,017,572. 2000.
6. Meyer R.S.; Cooper,K.L; Knorr,D; Lelieveld,H. High-pressure sterilization of foods. Food Technol.2000,54(11),67.
7. Knorr,D; Heinz,V. Development of nonthermal methods for microbial control.in Block,S.S. Disinfection, Sterilization and Preservation.Lippincott Williams & Wilkins.Philadelphia 2001,p.853.
8. Ludikhuyze,L.R; Hendrickx,M. The effect of high hydrostatic pressure on food quality related enzymes. in.Hendrickx,M; Knorr,D. High Pressure Treatment of Foods, Aspen Publishers, Inc. Gaithersburg, MD (in press).

9. Pfister,M.K.H; Butz,P;Heinz,V; Dehne,L.I; Knorr,D; Tauscher,B. Der Einfluss der Hochdruckbehandlung auf chemische Veränderungen in Lebensmitteln. BgVV, Berlin.2000

10. Dörnenburg,H; Knorr, D. Impact of physical stressors on food quality related enzymes. (unpublished data)

11. Dörnenburg,H; Knorr,D. Monitoring the impact of high-pressure processing on the biosynthesis of plant metabolites using plant cell cultures. Trends Food Sci.Technol. 1998,9,355-367.

12. Kasai,M; Hatae,K.; Shimada,A. Pressure induced hardening of vegetables and ist mechanisms. Int.Conference High Pressure Biosci.Biotechnol. Kyoto. 1995. p.109.

13. Kasai,M;Hatae,K; Shimada,A; Ibuchi,S. Pressure pretreatment of vegetables for controlling hardness before cooking. Nippon Shokuhin Kagaku Kaishi.1995.42.594-601.

14. Stute,R; Kluingler,R.W; Boguslawski,S; Eshtiaghi,M,N; Knorr,D. Effects of high pressure treatments on starches. Starch,1996,48,399-408.

15. Basak, S; Ramaswamy,H.S. Effect of high pressure processing on the texture of selected fruits and vegetables. J.Texture Stud. 1998,29,578-601.

16. Michel,M; Autio,K. High pressure - a tool for food structure engineering.in. Hendrickx ,M; Knorr,D. High Pressure Treatment of Foods. Aspen Publishers, Inc., Gaithersburg,MD (in press).

17. Teichgräber, P.; Knorr,D Effect of high pressure on membrane related enzymes of cultured plant cells. (unpublished data).

18. Koch,H.; Knorr,D. Effect of high pressure on plant foods. unpublished data

19. Bakki, R.; Knorr,D. Formation of heat sterilizable fruit gels through high pressure treatment. (unpublished data)

20. Cheftel,J.C.; Culioli,J. Effects of high pressure on meat: a review. Meat Sci.1997, 46,211-236.

21. Amanatidou,A; Schlüter,O.; Lemkau,K; Gorris,L.G.M, Smid,E.J.; Knorr,D. Effect of combined application of high pressure treatment and modified atmospheres on the shelf life of fresh Atlantic salmon. Innovative Food Sci. & Emerging Technol. 2000,1,87-98.

22. Lee,D.U; Heinz,V; Knorr,D. High pressure treatment of liquid whole eggs. (unpublished data).

23. Ludikhuyze,L; Hendrickx,M. The effect of high pressure on quality related chemical aspects.in. Hendrickx,M; Knorr,D. High Pressure Treatment of Foods. Aspen Publishers, Inc. Gaithersburgh, MD (in press).

24. Bignon,J.Cold pasteurizer Hyperrbar for the sterilization of freah fruit juices. Fruit processing, 1996,2,46-48.

25. Donsi,G.; Ferrari,g; di Matteo,M. High pressure stabilization of orange juice: evaluation of the effects of process conditions. Ital. J. Food Sci.1996, 2, 99-106.

32

26. Watanabe,M; Arai,E; Kumeno,K; Homma,K. A new method for for producing non-heated jam sample: The use of freeze concentration and high pressure sterilization. Biol. Chem. 1991, 55, 2175-2176.

27. Kimura,K. Ida, M; Yoshida,Y; Ohki,K; Fukumoto,T; Sakui,N. Comparison of keeping quality between pressure-processed and heat processed jam: changes in flavor components, hue and nutritional elements during storage. Biosci. Biotechnol. Biochem. 1994, 58, 1386- 1391.

28. Koch,H; Knorr,D. Storage time dependent changes of pressure treated peas and broccoli. (unpublished data).

29. Eshtiaghi,M,N; Knorr, D. Potato cubes response to water blanching and high hydrostatic pressure. J. Food Sci. 1993, 58, 1371-1374.

30. Butz,P; Koller, D; Tauscher,B. Ultra-high pressure processing of onions: chemical and sensory changes. Lebensm. Wiss.Technol. 1994, 27, 463-46731.

31. Poretta,S. New finding on tomato products. Fruit Processing, 1996, 2, 58-65.

32. Knorr,D. Novel approaches in food-processing technology: new technologies for preserving foods and modifying function. Current Opinion Biotechnol. 1999,10, 485-491.

33. Angersbach, A; Heinz,V; Knorr, D: Effects of pulsed electric fields on cell membranes in real food systems. Innovative Food Sci. & Emerging Technol. 2000, 1, 135-149

34. Ho,S.Y; Mittal,G: S. Electroporation of cell membranes: a review. Crit. Rev. Biotechnol. 1996, 16, 349-362.

35. Barsotti, L; Cheftel, J.C. Food processing by pulsed electric fields.II. Biological aspects. Food Rev. Internat. 1999,15, 181-213.

36. Phillips,S; Heinz,V; Knorr,D. Effect of high electric field pulses on the germination behavior of *Bacillus subtilis* spores. (unpublished data).

37. Knorr, D; Angersbach, A. Impact of high-intensity electric field pulses on plant membrane permeabilization. Trends Food Sci. Technol. 1998, 9, 185-191.

38. Mönch,S., Heinz, V; Knorr, D. Development of a biorector for field pulse application on plant systems. (unpublished data).

39. Schreck, S; Dörnenburg, H; Knorr,D. Evaluation of hydrogen peroxide production in tomato (*Lycopersicon esculentum*) suspension cultures as a stress reaction to high pressure treatment. Food Biotechnol. 1996, 10, 163-171.

40. Knorr, D; Geulen, M; Grahl, T; Sitzmann, W. Food application of high electric field pulses. Trends Food Sci. Technol. 1994, 5, 71-75.

41. Eshtiaghi, M.N; Knorr,D. Anwendung elektrischer Hochspannungsimpulse zum Zellaufschluss bei der Saftgewinnung am Beispiel von Weintrauben. Lebensmittelverfahrenstechnik, 2000, 45, 23-27.

42. Eshtiaghi, M, N; Knorr, D. Impact of high electric field pulses on cell disintegration of plant foods. 2: Apple. Lebensmittelverfahrenstechnik (in press)

43. Eshtiaghi, M.N; Knorr, D. Impact of high electric field pulses on cell disintegration of plant foods. 3: Black-currants. Lebensmittelverfahrenstechnik (in press).

44. Watzke, H. J. Impact of processing on bioavailability examples of minerals in foods. Trends Food Sci. Technol. 1998, 9, 320-327.

45. Ade-Omowaye, B.I.O.; Angersbach, A. Eshtiaghi, M,N; Knorr, D. Impact of high intensity electric field pulses on cell permeabilization and as pre-processing step in coconut processing. Innovative Food Sci. & Emerging Technol. (in press).

46. Ade-Omowaye, B. I. O;, Angersbach, A; Taiwo, K.A; Knorr, D. The use of high intensity electric field pulses in producing juice from paprika (*Capsicum annuum* L.) (submitted).

47. Angersbach, A; Knorr, D. Application of electric field pulses as pretreatment for influencing dehydration characteristics and rehydration properties of potato cubes. Nahrung/Food 1997, 41, 194-200.

48. Eshtiaghi, M.N; Knorr, D. Method for treating sugar beet. Internat. patent WO 99/64634, 1999.

Chapter 3

Change in Radical-Scavenging Activity of Spices and Vegetables during Cooking

Hitoshi Takamura[1], Tomoko Yamaguchi[1], Junji Terao[2], and Teruyoshi Matoba[1,*]

[1]Department of Food Science and Nutrition, Nara Women's University, Kitauoya-nishimachi, Nara 630–8506, Japan
[2]Department of Nutrition, School of Medicine, The University of Tokushima, Kuramoto-cho 3, Tokushima 770–8503, Japan

The change in radical-scavenging activity of vegetables and spices during cooking was analyzed by the 1,1-diphenyl-2-picrylhydrazyl-HPLC method. After boiling of vegetables, the radical-scavenging activity increased in burdock, green pepper, asparagus, eggplant, and carrot, while the activity in other vegetables decreased. In most cases, however, the total activity of cooked vegetable and cooking water was higher than that of fresh vegetables. Microwave heating increased the radical-scavenging activity in 9 of 14 vegetables. These results suggest that the intake of vegetables with cooking water or microwave-cooked vegetables can be recommended to use radical-scavenging components efficiently. In the process of curry cooking, the radical-scavenging activity of spices decreased, while that of vegetables increased. The radical-scavenging activity of curry also decreased after heating.

© 2002 American Chemical Society

Free radicals and active oxygen species are well known to induce many types of oxidative damage to biomolecules, which cause cancer, aging, and life-style related diseases (*1*). Dietary foods contain a wide variety of free radical-scavenging antioxidants, such as flavonoids, vitamin C, and so on (*2*). These compounds are abundantly present in vegetables, fruits, spices, and other plant origin foods. Epidemiological studies have shown that higher intake of fresh vegetables, fruits, tea, and wine is associated with lower risk of cancer and life-style related diseases (*3,4*). Hence, there is currently much interest in natural antioxidants and their role in human health and nutrition. However, we usually eat foods after cooking. In this work, we demonstrated the changes in radical-scavenging activity of vegetables and spices during cooking.

Materials and Methods

Preparation of Fresh and Cooked Vegetable Extracts

Fourteen vegetables were obtained from local markets in Nara, Japan. The edible portions of fresh vegetables (10-20 g) were cut into small pieces and homogenized with 30 mL of water for 20-30 s, using a homogenizer. The resulting homogenate was centrifuged at 27,000×g for 20 min at 4°C, and the supernatant was filtered through a 0.45-μm filter. After appropriate dilution, the aqueous extract was used for the measurement of radical-scavenging activity. For the measurement of ascorbic acid content, 5% metaphosphoric acid with or without 1% stannous chloride was used instead of water.

For boiling of vegetables, the edible portions of fresh vegetables (80-150 g) were cut into small pieces. Cut vegetable was heated in 500 ml of boiling water for 5 min. Then, the extract of cooked vegetable was prepared as described above. The cooking water was also used for the measurement of radical-scavenging activity and ascorbic acid content after concentration.

For microwave heating, cut vegetable was placed on a ceramic plate, covered with a plastic cap, and heated in a microwave oven (500 W) for 5 min. Then, the extract of cooked vegetable was prepared as described above.

Curry Preparation

Curry is one of the most popular spicy dishes in Japan and other Asian countries which uses both vegetables and spices. Vegetables and meat were cooked in water, and kept separately. Spice mixture was heated with butter and wheat flour to prepare curry paste. Then, cooked vegetables and meat and curry paste were combined and heated to prepare curry.

Spice mixture used for curry cooking contained 9 flavoring spices; coriander (27%), cumin (8%), cardamon (5%), allspice (4%), cinnamon (4%), fenugreek (4%), clove (2%), fennel (2%), and mace (2%), 3 hot-tasting spices; ginger (4%), red pepper (4%), and white pepper (4%), and 1 coloring spice; turmeric (30%). This is a Japanese recipe for a popular curry powder with medium hot taste (5). One serving (260 g) of curry was prepared from 30 g carrot, 40 g onion, 60 g potato, 50 g beef, 2.5 g spice mixture, 12.5 g butter, and 10 g wheat flour.

The curry preparation at each step was first extracted with 80% ethanol, and then with water. Extraction was carried out under the same procedure for fresh vegetables as described above. Both 80% ethanol and water extracts were used for the determination of radical-scavenging activity.

Determination of Radical-scavenging Activity and Ascorbic Acid Content

The radical-scavenging activity of ascorbic and dehydroascorbic acids, vegetables, and curry preparations was determined by 1,1-diphenyl-2-picrylhydrazyl (DPPH)-HPLC method according to Yamaguchi *et al.* (*6*). Briefly, 1 mL of 0.5 mM DPPH in ethanol was mixed with 0.8 mL of 100 mM Tris-HCl buffer (pH 7.4) and 0.2 mL of sample extract. The mixture was shaken vigorously and left to stand for 20 min at room temperature in the dark. Then, the reaction mixture was injected into HPLC with a 20-µl loop. HPLC analysis was performed using a TSKgel Octyl-80Ts column (4.6×150 mm, Tosoh, Tokyo, Japan) at room temperature with a mobile phase of methanol/water (70:30, v/v), and at a flow rate of 1 mL/min. DPPH was detected at 517 nm. Trolox was used as the standard to evaluate radical-scavenging activity. The ascorbic acid content was determined by HPLC according to Kishida *et al.* (*7*). Then, the radical-scavenging activity due to ascorbic acid was calculated.

Results and Discussion

Radical-scavenging Activity of Ascorbic and Dehydroascorbic Acids

The radical-scavenging activity of ascorbic and dehydroascorbic acids was determined by the DPPH-HPLC method (Table I). Dehydroascorbic acid (50 µM) had no ability to scavenge DPPH radical, while the activity of 50 µM ascorbic acid was approximately 60 μM Trolox equivalent. Electron spin resonance measurement also showed that dehydroascorbic acid has no scavenging activity on hydroxyl radical (data not shown). The combined effects of ascorbic acid and dehydroascorbic acid on DPPH radical-scavenging activity were also measured (Table I). There was no positive or negative synergism between the radical-scavenging activity of ascorbic and dehydroascorbic acids.

Change in Radical-scavenging Activity of Vegetables During Boiling

Figure 1 shows the radical-scavenging activity of fresh vegetables, boiled vegetables, and cooking water. In burdock, green pepper, asparagus, eggplant, and carrot, the radical-scavenging activity of cooked tissue was higher than that of fresh tissue, though the ascorbic acid content decreased. This increase may be due to the thermal destruction of vegetable cell walls and subcellular compartments, the decrease of oxidative loss of active compounds by inactivation of oxidase, and/or the activation of inactive compounds by heating.

In broccoli, pumpkin, and cabbage, the radical-scavenging activity in the tissue decreased after boiling. However, part of the activity was found in cooking water, and the total activity of cooked tissue and cooking water increased. Hence, in order to intake the radical-scavenging components efficiently, intake of cooking water together with cooked vegetable can be recommended, such as vegetable soup and stew.

In Chinese cabbage, onion, kidney beans, Japanese radish, tomato, and spinach, the total radical-scavenging activity of cooked tissue and cooking water decreased after boiling. The active components in these vegetables may be unstable against heat.

Effect of Sodium Chloride on the Radical-scavenging Activity of Boiled Vegetable

Table II shows the radical-scavenging activity of boiled spinach with or without sodium chloride. As the content of added sodium chloride increased, the radical-scavenging activity was retained more in the tissue, while the total activity did not change. Therefore, the addition of sodium chloride to the cooking water can be recommended to intake the radical-scavenging components efficiently, since it inhibits the release of the active components from the tissue to the cooking water.

Table I. Radical-scavenging Activity of Ascorbic and Dehydroascorbic Acids

Concentrations (μM)		Radical-scavenging Activity
Ascorbic Acid	Dehydroascorbic Acid	(μM Trolox equivalent)
0	50	ND[a]
50	0	56.9±1.9[b]
50	25	59.5±0.4
50	50	59.2±1.5
50	75	59.9±2.6
50	100	61.0±2.1
50	200	59.9±1.3
50	250	60.0±1.9
50	500	58.8±0.8

[a]ND, not detected.

[b]The values are the means±SD of 3 determinations.

Table II. Effect of Sodium Chloride on Radical-scavenging Activity of Boiled Spinach

Concen-tration	Radical-scavenging Activity		(μmol Trolox equivalent/100 g)	
	Total Activity		Ascorbic Acid[a]	
	Cooked Tissue	Cooking Water	Cooked Tissue	Cooking Water
0	73.6[b] (20)[c]	284.4 (76)	9.1 (10)	6.3 (7)
1	92.0 (25)	217.2 (58)	6.6 (7)	1.3 (1)
2	156.7 (42)	154.1 (41)	20.9 (23)	8.2 (9)

[a]Calculated from ascorbic acid content.

[b]The values are the means of 3 determinations.

[c]The values in parentheses are the percentages to the activities of fresh vegetables.

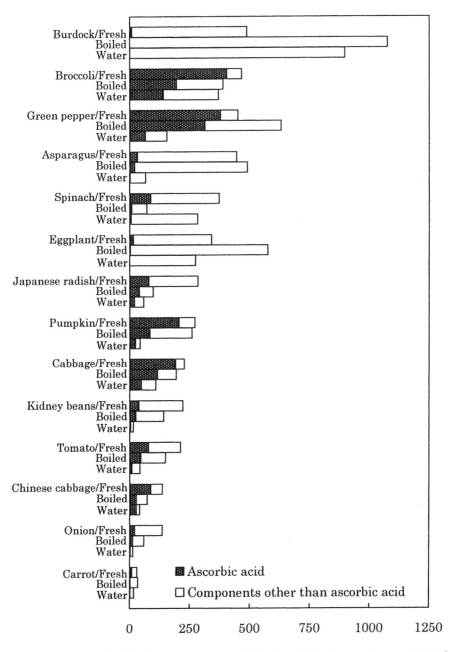

Figure 1. Change in radical-scavenging activity of vegetables during boiling.

Change in Radical-scavenging Activity of Vegetables During Microwave Heating

Figure 2 shows the radical-scavenging activity of fresh and microwave-heated vegetables. In 9 of 14 vegetables, especially in burdock and eggplant, the radical-scavenging activity increased by microwave heating. The activities of microwave-heated vegetables were higher than those of boiled vegetables, since there was no release of active components to boiling water during microwave heating. Hence, microwave heating can be recommended to intake the radical-scavenging components efficiently. In onion, Japanese radish, spinach, kidney beans, and pumpkin, the radical-scavenging activity decreased after microwave heating. However, microwave heating is effective in the cooking of spinach, since the loss of the activity was minimized by microwave heating.

Change in Radical-scavenging Activity of Vegetables and Spices During Curry Cooking

The changes of the radical-scavenging activity during curry cooking are shown in Figure 3. The activity of fresh vegetables and meat was 75 μmol Trolox equivalent per one serving. The activity increased to 205 μmol after heating. On the contrary, the activity of spice mixture decreased from 318 to 162 μmol during preparation of curry paste. This decrease may be due to decomposition or evaporation of active compounds, since spices were heated with butter at high temperature. In spite of the loss by heating, however, spices still contained high activity.

The activity of curry was 312 μmol Trolox equivalent per one serving, which is 20% lower than the total activity (367 μmol) of cooked vegetables and meat and curry paste. Hence, heating of curry for a long time cannot be recommended to keep radical-scavenging activity. However, spicy dishes such as curry are useful to intake radical-scavenging components contained in spices as well as vegetables.

References

1. Halliwell, B.; Gutteridge, J. M. C.; Cross, C. E. *J. Lab Clin. Med.* **1970**, *119*, 598-620.
2. Shahidi, F; Naczk, M. *Food Phenolics, Sources, Chemistry, Effects, Applications;* Technomic Publishing Co. Inc.: Lancaster, PA, 1995; pp 75-107.

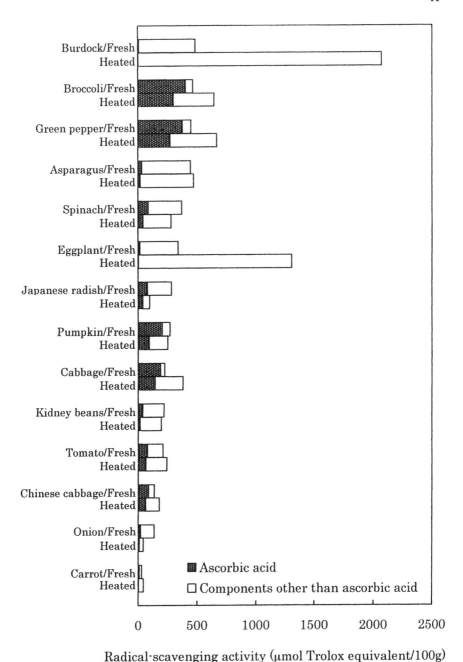

Figure 2. Change in radical-scavenging activity of vegetables during microwave heating.

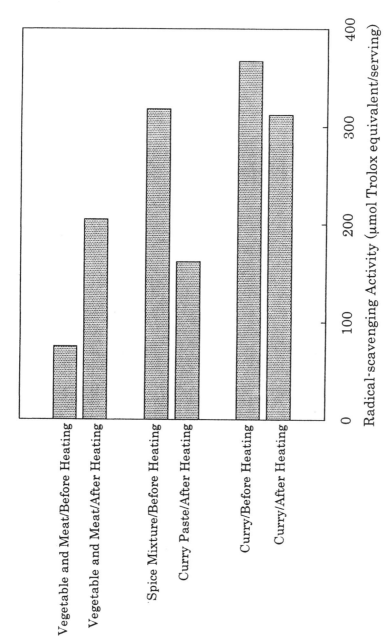

Figure 3. Change in radical-scavenging activity of vegetables and spices during curry cooking.

3. Ames, B. N.; Shigenaga, M. K.; Hagen, T. M. *Proc. Natl. Acad. Sci. U.S.A.* **1993**, *90*, 7915-7922.
4. Willett, W. C. *Science*, **1994**, *264*, 532-537.
5. Hamada, T. *Jpn. Food Sci.*, **1970**, *12*, 51-56.
6. Yamaguchi, T.; Takamura, H.; Matoba, T.; Terao, J. *Biosci. Biotechnol. Biochem.* **1998**, *62*, 1201-1204.
7. Kishida, E.; Nishimoto, Y.; Kojo, S. *Anal. Chem.* **1992**, *64*, 1505-1507.

Chapter 4

In Vitro Biogeneration and Stability of Pure Thiosulfinates from Alliums: Stability and Reactivity of Thiosulfinates

Cunxi Shen and Kirk L. Parkin*

Department of Food Science, University of Wisconsin,
Babcock Hall, 1605 Linden Drive, Madison, WI 53706

A model reaction system was developed for preparing pure homologous thiosulfinates and propanethial-S-oxide (PTSO) using an isolated alliinase and alk(en)yl-L-cysteine sulfoxides (ACSO). Thiosulfinates were characterized for their intrinsic pH and thermal sensitivities. Thiosulfinate decay could be fitted to first-order processes over the pH range of 1.2-9.0 and at 20-80°C. The pH dependence of thiosulfinate stability in descending order was: pH 4.5-5.5 > pH 1.2 > pH 6.5-7.5 > pH 8.0-9.0 in the presence of 0.1 M Tris. Thiosulfinates with propyl and ethyl substituent alk(en)yl groups were generally more stable than those with allyl and methyl groups. Thiosulfinates underwent alkyl-exchange reactions at pH 8-9 without net loss of total thiosulfinate levels within 60-90 min at 20°C.

Organosulfur components in various tissue preparations of *Allium* species have been of interest for decades. Epidemeological data and demonstrations (primarily, *in vitro*) of biological activity of various tissues extracts, pastes and in some cases even pure organosulfur compounds have sustained a widely held belief that these compounds possess therapeutic potential in humans (*1,2*). The precursors of these organosulfur compounds in intact *Allium* tissues are *S*-alk(en)yl-L-cysteine sulfoxides (ACSO), which upon tissue disruption, are acted upon by an endogenous alliinase (E.C. 4.4.1.4) to yield the sulfenic acid (RSOH) scission product. The latter either condenses (when R = methyl, propyl, 1-propenyl and 2-propenyl) to form homologous (R_1 = R_2) or heterologous ($R_1 \neq R_2$) thiosulfinates ($R_1S(O)S\ R_2$), or rearrange to form propanethial-*S*-oxide (PTSO) when R = 1-propenyl as depicted in Scheme 1.

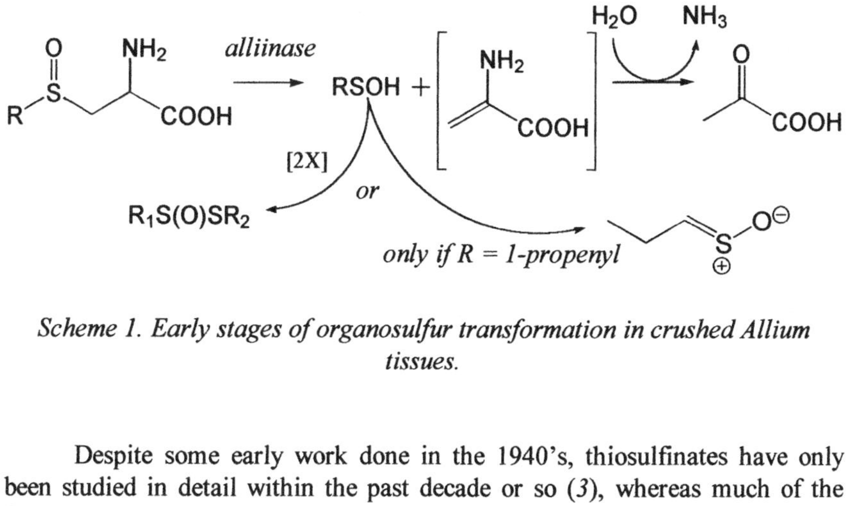

Scheme 1. *Early stages of organosulfur transformation in crushed Allium tissues.*

Despite some early work done in the 1940's, thiosulfinates have only been studied in detail within the past decade or so (*3*), whereas much of the focus in previous studies on organosulfur compounds of Alliums had been placed on the various sulfides ($R_1(S)_xR_2$) (*1,2*). It is now evident that thiosulfinates may possess or mediate many biological activities, including flavoring (*4,5*), antioxidant (*6*), antimicrobial (*7,8*), virucidal (*9*), antithrombotic (*10,11*), antiasthmatic (*12*) and anticarcinogenic (*13,14*) activities. However, what is lacking in these areas of study is a knowledge of structure-function relationships among thiosulfinates and their attendant biological activities. Furthermore, some studies have attributed certain biological effects of the use of pastes, distilled oils or extracts of *Allium* tissues to thiosulfinates (*15,16*), despite the presence of a plethora of other components that may be the active principle(s).

To address the limitations just cited, it became evident to us that having a facile and relatively abundant source of pure thiosulfinates would facilitate next generation studies focused on establishing structural determinants of various biological and chemical properties of thiosulfinates. Furthermore, since foods are the natural vehicles for delivering *Allium*-based organosulfur compounds to the public, it was also evident that understanding the intrinsic stabilities of thiosulfinates to conditions prevailing, or that could be encountered, during food handling, processing or distribution would be of importance in understanding the fate of these compounds in their host tissues. Understanding the chemistry of these organosulfur compounds in tissue preparations is complicated by the transient nature of thiosulfinates (*4,17*) and the potential reactivity with many other intrinsic components.

Thiosulfinates and related compounds have been chemically synthesized or isolated from *Allium* tissues (*3,19*). We sought to develop an alternative process for the preparation of pure thiosulfinate species and PTSO found in freshly prepared *Allium* extracts. This process was based on reacting ACSO and an isolated onion alliinase preparation in a system that simulates the biogeneration of these compounds *in situ*.

Materials and Methods

Materials

Chemicals were obtained from Sigma (St. Louis, MO) or Aldrich (Milwaukee, WI) Chemical Companies unless otherwise noted. All solvents used were chromatography grade. White onion bulbs were purchased from a local retail market.

Preparation of Immobilized Alliinase

A crude alliinase preparation was prepared using the steps of homogenization, 65% saturated ammonium sulfate fractionation and dialysis at 0-4°C as described earlier (*19*). Immoblization of crude alliinase in alginate gels was also carried out as described earlier (*19*). The resulting enzyme-loaded gel was sliced into 5-10 mm thick slices and stored at 4°C until used.

Preparation of S-Alk(en)yl-L-cysteine Sulfoxide (ACSO) Substrates

Diasteromeric MCSO was prepared essentially by the method of Synge and Wood (*20*) with yields of 60-80%. Synthesis of diasteromeric ECSO was as described for MCSO except that *S*-ethyl-L-cysteine was used as the starting material, and yields were 60-75%. Synthesis of diasteromeric PCSO was similar to the method described by Lancaster and Kelly (*21*), and yields were 45-55%. Synthesis of diasteromeric 2-PeCSO was similar to that described for PCSO except that 2-propenyl bromide was substituted for propyl bromide, and yields were about 15-20%. (+)*S*-1-Propenyl-L-cysteine sulfoxide (1-PeCSO) was isolated from white onion bulbs using a modified method of Carson et al. (*22*), with yields of about 1 g from 5-6 kg tissue. All of these methods are described in detail in an upcoming paper (*23*).

Model Reaction System

Typically, ACSO (0.025 mmol 1-PeCSO; 0.05 mmol for all others) was combined with 0.5 g immobilized alliinase (ground to yield gel fragments of $\leq$ 1 mm dimension) in 4.0 ml of 0.1 M Tris (pH 7.5) at 21-23°C. After pre-determined intervals of incubation, thiosulfinate ($R_1S(O)SR_2$; where R_1/R_2 groups are methyl (Me), ethyl (Et), propyl (Pr), or allyl (All); 1-propenyl species were not evaluated in this study) and/or PTSO products were obtained by extracting an aqueous sub-sample of the reaction mixture into an equal volume of $CHCl_3$ (containing benzyl alcohol as internal standard).

Quantification of organosulfur analytes was based on the peak area in normal phase HPLC chromatography on a 250 mm × 4.6 mm, Microsorb 5 μm Silica column (Rainin Instrument Co. Inc., Woburn, MA) using gradient elution with 2-propanol:hexane from 2:98 (v/v) held for 6 min to 10:90 (v/v) for the next 10 min followed by a 7 min hold, with an elution rate of 1.4 ml min^{-1} (adapted from reference *4*). HPLC molar response factors were based on quantification of each thiosulfinate or PTSO component in a stock solution as determined by their ^1H NMR signals in comparison to the signal of the internal standard (*ter*-butanol). Corrections were also made for the efficiency of $CHCl_3$ extraction, where necessary.

The progress of model reactions was also followed by pyruvate formation, based on the coupling assay involving oxidation of NADH to NAD by lactate dehydrogenase (*24*).

For all these studies, at least two duplicate experiments were conducted and mean values are reported, where CV ranged 3-15% among the studies.

Structures of individual ACSO, and the thiosulfinates prepared therefrom, were confirmed by ^{1}H NMR (model AM-300 NMR spectrometer, Bruker Instruments, Inc., Billerica, MA) operated at 300 MHz, using D_2O or $CDCl_3$ in the sample diluent.

Thiosulfinate Stability and Reactivity Studies

Pure thiosulfinates (0.35-3.5 mM) were dissolved in aqueous 0.1 M Tris at 20°C and four subsamples were prepared at different pH values of 1.2, 5.5, 7.5 and 9.0 by addition of dilute NaOH or HCl. It is acknowledged that Tris is an effective buffer only in the pH range of about 7.3-9.1 at 20°C. The primary reason it was included was that it is an appropriate buffer for studies on onion alliinase, as well as in our model reaction systems, because of the slightly alkaline pH optimum for alliinase (*19,25,26*). Immediately after pH adjustment, each subsample was analyzed for thiosulfinate level at "zero-time" by HPLC. Analysis for thiosulfinate level was repeated at pre-determined intervals during the course of incubation under the conditions being evaluated. First-order rate constants for decay were calculated as the slopes of Ln[thiosulfinate] *versus* time plots, and these constants were transformed into half-lives. Regression analysis for these semi-log plots used to determine decay rate constants proved to be highly linear fits ($r^2 \geq 0.96$).

No attempt was made to account for the ΔK_a of -0.031 °C^{-1} for Tris buffer when thiosulfinate solutions were incubated at different temperatures, as this was judged to be impractical. For example, a thiosulfinate-containing Tris solution to be examined at 80°C at pH 9.0 would have to be initially prepared at 20°C at pH ~10.8. This is outside of the pH buffering range of Tris, making appropriate pH adjustment subject to uncertainty. Moreover, a pH of near 11 is also a condition where thiosulfinate decay could be rapid enough (*2,27*) to cause thiosulfinate decay in the time taken to simply prepare the sample for incubations under the specific conditions to be evaluated (see also results and discussion). Alternatively, all samples were prepared at 20°C at pH 1.2, 5.5, 7.5 or 9.5 in 0.1 M Tris, and then incubated at various temperatures (actual pH values were measured and recorded at each temperature). As a result, different (but similarly spaced intervals of) pH values were used to examine thiosulfinate decay at different temperatures.

An evaluation of the temperature dependence of decay for each thiosulfinate was done by plotting the estimated first-order rate constants on Arrhenius plots, and estimating an E_a value from the slope of these plots. Because pH values were not constant over the range of temperatures used for each of these plots, the E_a values obtained were referred to as "pseudo-E_a" values, and these may differ from actual E_a values.

Results and Discussion

Progress of Model Reaction Mixtures

A typical progress curve for a model reaction under "standard" conditions is shown in Figure 1. Under these conditions (0.1 M Tris, pH 7.5), the relative rate of pyruvate formation observed for ACSO substrates were: 1-PeCSO > 2-PeCSO > PCSO > ECSO > MCSO. This pattern is consistent with the relative reactivity of ACSOs and onion alliinase as predicted from respective K_m values for these substrates (19,25,26). Transformation of the data in Figure 1 to linear (log-log) plots (not shown) yielded slopes ($r^2 \geq 0.94$) that correspond to relative V_m/K_m values (viz., selectivity constant - theory in reference 28). Relative values for V_m/K_m for MCSO, ECSO, PCSO, 2-PeCSO and 1-PeCSO substrates were: 1.00:1.59:2.01:4.38:10.7, consistent with respective V_m/K_m values of 1:3:18 previously determined for MCSO:PCSO:1-PeCSO (calculated from data in reference 26).

The patterns for thiosulfinate formation in reaction mixtures (Figure 2; corrected for extraction efficiency) were consistent with the patterns of pyruvate evolution (Figure 1), in that the same order of preferences of ACSO reactivity were observed (except for 1-PeCSO). Based on the anticipated reaction stoichiometry of 2:1 pyruvate:thiosulfinate formed (Scheme 1), pyruvate levels were only 36-104% of those expected based on thiosulfinate yields. This pyruvate deficit or "crypto-pyruvate" phenomenom has been reported previously (29), and an explanation for this remains enigmatic.

Another anomaly between the trends in Figures 1 and 2 is for reactions employing 1-PeCSO as substrate, where limited levels of PTSO accumulated in view of the levels of pyruvate evolved. Subsequent experiments revealed that PTSO accumulation after 2 hr incubation reached 0.1 mM for reactions conducted in sodium phosphate-buffered systems at pH 5.0 compared to non-detectable levels in Tris-buffered systems at pH 7.5. Greatest levels of PTSO accumulation observed were 0.2 mM and this was within 2-4 min reaction in either Tris- or phosphate-buffered systems at pH 7.5. These observations

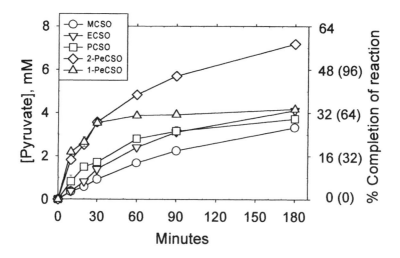

Figure 1. Pyruvate generation in model reaction mixtures. Reaction mixtures contained 0,05 mmol of either MCSO, ECSO, PCSO, or 2-PeCSO, or 0.025 mmol 1-PeCSO, and alliinase in Tris (7.5) buffer. Right ordinate denotes % reaction yield (in parentheses for 1-PeCSO) assuming a stoichiometric production of pyruvate (Figure also appears in reference 23).

indicated the sensitivity of PTSO to further transformation (or decay) at near-neutral pH conditions.

These reactions systems yielded pure (99%) thiosulfinates, as confirmed by ^{1}H-NMR analysis. For routine preparation of pure thiosulfinates for studies of intrinsic chemical properties and stabilities, typically, 200 mg ACSO was reacted with alliinase in Tris-buffered (pH 7.5) medium for 2-3 hr. The thiosulfinate product was extracted into $CHCl_3$, the solvent was immediately evaporated under vacuum (~80 mm Hg) at 20-22°C, and the residue was reconstituted in water, with a final product yield determined by HPLC to be about 60-80 μmoles. This model reaction system can also be used to prepare homologous and heterologous thiosulfinates from binary ACSO substrate systems (23). To prepare PTSO, extraction of the reaction mixture after only 2-10 min of incubation yielded the greatest levels of recoverable PTSO.

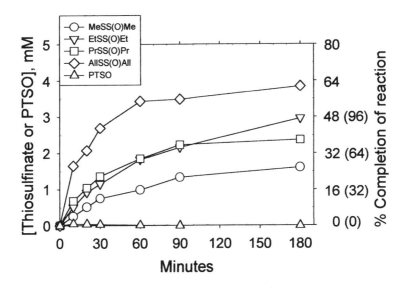

Figure 2. Organosulfur product generation in model reaction mixtures. Legend is the same as for Figure 1, except that products formed were thiosulfinates, RS(O)SR, where R = methyl (Me), ethyl (Et), propyl (Pr), 2-propenyl (All), or propanethial-S-oxide (PTSO), from the corresponding ACSO (Figure also appears in reference 23).

Thiosulfinate Stability as a Function of pH

Stability of the homologous thiosulfinates tested was dependent on pH, with a descending order of stability at pH: 5.5 > 1.2 > 7.5 > 9.0 in the presence of Tris (Table I).

Thus, slightly acidic media were most conducive to stabilizing thiosulfinates, as indicated by an earlier report (27). The present results are also consistent with the reported and relative stabilities of allicin (AllS(O)SAll) and methyl methanethiosulfinate in aqueous medium at 23°C (27). What was most surprising was just how stable some thiosulfinates were, such as the complete retention of PrS(O)SPr and EtS(O)SEt levels for 4 mo at 20°C and pH 5.5, in view of the reported *in situ* instability of various thiosulfinates (4,17,30). Our results also indicated a correlation of stability with longer and saturated substituent alk(en)yl substituent groups, and the relative stabilities of homologous thiosulfinates over the pH range of 1.2-9.0 at 20°C were: PrS(O)SPr ~ EtS(O)SEt > MeS(O)SMe > AllS(O)SAll. Another interesting observation was that PrS(O)SPr was more stable than EtS(O)SEt at pH 7.5-9.0, but the reverse was true at pH 1.2. This relationship was retained throughout the full range of temperatures studied (see below).

Table I. Half-lives of Thiosulfinates in Aqueous 0.1 M Tris at 20°C.

$pH \rightarrow$ *Thiosulfinate* $\downarrow$	1.2	5.5	7.5	9.0
AllS(O)SAll	20.7 days	52.3 days	3.2 days	5.8 hr
PrS(O)SPr	180 "	stable*	25.4 "	69.5 "
EtS(O)SEt	282 "	stable*	19.0 "	22.3 "
MeS(O)SMe	69.6 "	179 "	4.9 "	13.0 "

Half-lives were determined from semi-log plots of thiosulfinate decay and corresponding first order rate constants (k^{-1}). *Stable indicates that there was no evidence of decay after 4 mo.

As might be expected from Table I, attempts to study thiosulfinate stability in aqueous media at 4°C were made difficult by the inability to observe decay within a convenient time frame (preempting reliable estimates of half-lives). Instead, thiosulfinate stability was assessed at elevated temperatures in aqueous media containing 0.1 M Tris. At 40°C, the pattern of pH-dependence and relative stabilities of thiosulfinates observed (Table II) was identical to that observed at 20°C. At 60°C (Table III) and 80°C (Table IV), a pattern emerged that was similar to those observed at the lower temperatures studied, with one exception. MeS(O)SMe became the most stable thiosulfinate under acidic conditions (pH 1.2-4.5) between 60-80°C.

Table II. Half-lives of Thiosulfinates in Aqueous 0.1 M Tris at 40°C.

$pH \rightarrow$ *Thiosulfinate* $\downarrow$	1.2	5.1	7.1	8.8
AllS(O)SAll	1.7 days	2.7 days	0.57 days	1.0 hr
PrS(O)SPr	11.0 "	97.2 "	6.2 "	9.9 "
EtS(O)SEt	14.5 "	78.8 "	4.6 "	5.6 "
MeS(O)SMe	9.2 "	32.9 "	1.3 "	1.8 "

Half-lives were determined from semi-log plots of thiosulfinate decay and corresponding first order rate constants (k^{-1}).

Table III. Half-lives of Thiosulfinates in Aqueous 0.1 M Tris at 60°C.

pH → Thiosulfinate ↓	1.2	4.8	6.8	8.4
AllS(O)SAll	5.0 hr	6.9 hr	2.5 hr	0.45 hr
PrS(O)SPr	45.2 "	355 "	19.9 "	3.8 "
EtS(O)SEt	66.0 "	254 "	15.2 "	2.6 "
MeS(O)SMe	30.2 "	170 "	5.4 "	0.70 "

Half-lives were determined from semi-log plots of thiosulfinate decay and corresponding first order rate constants (k^{-1}).

Table IV. Half-lives of Thiosulfinates in Aqueous 0.1 M Tris at 80°C.

pH → Thiosulfinate ↓	1.2	4.5	6.4	8.0
AllS(O)SAll	0.34 hr	0.82 hr	18.7 min	3.4 min
PrS(O)SPr	1.4 "	2.2 "	104 "	22.0 "
EtS(O)SEt	1.7 "	4.0 "	189 "	18.0 "
MeS(O)SMe	2.1 "	10.0 "	62.1 "	7.4 "

Half-lives were determined from semi-log plots of thiosulfinate decay and corresponding first order rate constants (k^{-1}).

The temperature-dependent shift in order of stability of homologous thiosulfinates prompted an estimation of "pseudo-E_a" values (Table V), despite the limitations previously identified in the experimental section. Considering that pH ranged as much as 1.0 unit for each interval tested, linear regression fits for Arrhenius-type plots were reasonable, at $r^2 \geq 0.97$ for AllS(O)SAll, $r^2 \geq 0.92$ for PrS(O)SPr, $r^2 \geq 0.93$ for EtS(O)SEt, $r^2 \geq 0.96$ for MeS(O)SMe for each of the four pH ranges evaluated for each thiosulfinate. Generally, "pseudo-E_a" values were greater in acidic media compared to alkaline media for all thiosulfinates, consistent with different mechanisms of decay under these conditions. A closer look at Table V indicates that the energetics of decay was similar between members within each of the thiosulfinate pairs of MeS(O)SMe and AllS(O)SAll, and PrS(O)SPr and EtS(O)SEt. Each member of the former pair exhibited similar "pseudo-E_a" values at both pH ranges of 1.2 and 4.5-5.5, whereas each member of the latter pair exhibited greater "pseudo-E_a" values at pH 4.5-5.5 than at pH 1.2. A final trend that was observed was that decay kinetics of MeS(O)SMe was the least temperature dependent of the thiosulfinate species tested.

Table V. Pseudo-activation Energies (E_a) for Thiosulfinate Decay in Aqueous 0.1 M Tris at 20-80°C.

pH → Thiosulfinate ↓	1.2	4.5-5.5	6.4-7.5	8.0-9.0
AllS(O)SAll	103	104	75.4	72.9
PrS(O)SPr	111	158	91.5	76.2
EtS(O)SEt	113	140	87.0	78.1
MeS(O)SMe	93.8	83.9	72.1	73.1

Pseudo-E_a values were determined from Arrhenius plots from the data in Tables 1-4. Values are expressed in kJ mol^{-1}.

Thiosulfinate Alk(en)yl-exchange Reactions

During preliminary studies on stability of binary homologous thiosulfinate systems under slightly alkaline conditions, the evolution of heterologous thiosulfinate species was observed. This prompted further studies into the phenomenon of alk(en)yl-exchange reactions in aqueous 0.1 M Tris systems, a process observed earlier for neat or benzene solutions of thiosulfinates (31). The pH dependence of alkyl-exchange between PrS(O)SPr and EtS(O)SEt indicated that net exchange was maximized at pH 9-10 after 90 min, and retention of total thiosulfinate levels was maximized at pH 8-9 (Figure 3).

Thus, there appeared to be a delicate balance of influence of pH on reactivity (alkyl-exchange) and net decay of thiosulfinates, and pH 9 was chosen as the optimum condition to follow the time course of exchange while maximizing retention of initial thiosulfinates levels (Figure 4). Within the 90 min period, a progressive loss in homologous thiosulfinate species was accounted for in a reciprocal evolution of heterologous thiosulfinate species, while total thiosulfinates were conserved at near-initial levels. These trends allow the suggestion that this alkyl-exchange or "scrambling" process is random and that the equilibrium molar distribution of heterologous:homologous species would ultimately reach 1:1, provided that no selective monomolecular decay of a particular species takes place during the course of incubation.

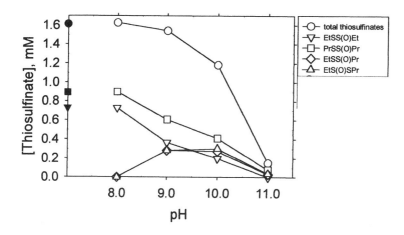

Figure 3. Alkyl-exchange and decay of thiosulfinates as a function of pH. PrS(O)SPr and EtS(O)SEt were initially included at the levels indicated by closed symbols on the ordinate. Resulting thiosulfinate profiles were analyzed after 90 min incubation at 20-23°C in 0.1 M Tris at the pH indicated.

It is revealing to juxtapose the patterns in Figure 4 with the half-lives of stability of PrS(O)SPr and EtS(O)SEt in 0.1 M Tris (pH 9.0) at 20°C (Table I). Although one can consider these thiosulfinates to be fairly stable under these conditions, they are not unreactive or chemically static. Despite their propensity for alkyl-exchange reactions, the basic thiosulfinate structure is conserved, apparently because competing or parallel reaction pathways leading to thiosulfinate decay are not favored under these conditions. Thus, one can easily envision this dynamic intermolecular exchange process taking place between two molecules of homologous thiosulfinates of the same species (although we can not directly observe this analytically).

Conclusions

An alliinase-based system for biogeneration of thiosulfinates and PTSO could be used to prepare pure thiosulfinates (or PTSO) for the study of intrinsic chemical properties of these organosulfur compounds. Thiosulfinate decay was characterized by first-order processes and the alk(en)yl substituent group(s) of the thiosulfinate appeared important in conferring the relative pH and thermal stability. Intrinsic stabilities of pure thiosulfinates *in vitro* were considerably greater than they appear to be *in situ*. Thus, while thiosulfinates are

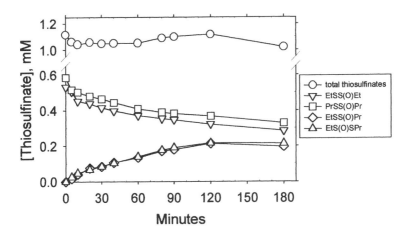

Figure 4. Progress of alkyl-exchange between homologous thiosulfinates at pH 9.0 in 0.1 M Tris at 20-23°C.

intrinsically stable, they remain reactive, and there are likely many chemical components within *Allium* tissue preparations that mediate thiosulfinate reactivity. Identifying these other chemical components will be important to developing strategies to control the fate of thiosulfinates in foods during, or as a consequence of, processing. Some effective processing strategies may prove to be rather simple, and it was demonstrated that brief pH adjustment could be used to modify the thiosulfinate profile while still preserving total thiosulfinate levels.

Acknowledgment

This work was supported by the College of Agricultural and Life Sciences of the University of Wisconsin-Madison, and The United States Department of Agriculture (Grants 96-35500-3352, 58-3148-7-031, and 97-36200-5189).

References

1. Kendler, B.S. *Prev. Med.* **1987**, *16*, 670-685.
2. Lau, B.H.S.; Tadi, P.P.; Tosk, J.M. *Nutr. Res.* **1990**, *10*, 937-948.
3. Block, E. *Angew. Che. Int. Ed. Engl.* **1992**, *31*, 1135-1178.

4. Block, E.; Naganathan, S.; Putnam, D.; Zhao, S-H. *J. Agric. Food Chem.* **1992**, *40*, 2418-2430.
5. Block, E.; Putnam, D.; Zhao, S-H. *J. Agric. Food Chem.* **1992**, *40*, 2431-2438.
6. Rabinkov, A.; Miron, T.; Konstantinovski, L.; Wilchek, M.; Mirelman, D.; Weiner, L. *Biochim. Biophys. Acta* **1998**, *1379*, 233-244.
7. Cavallito, C.J.; Bailey, J.H. *J. Am. Chem. Soc.* **1944**, *66*, 1950-1951.
8. Kyung, K.H.; Fleming, H.P. *J. Food Sci.* **1994**, *59*, 350-355.
9. Weber, N.D.; Andersen, D.O.; North, J.A.; Murray, B.K.; Lawson, L.D.; Hughes, B.G. *Planta Med.* **1992**, *58*, 417-423
10. Lawson, L.D.; Ransom, D.K.; Hughes, B.G. *Thromb. Res.* **1992**, *65*, 141-156.
11. Morimitsu, Y.; Morioka, Y.; Kawakishi, S. *J. Agric. Food Chem.* **1992**, *40*, 368-372.
12. Bayer, T.; Breu, W.; Seligmann, O.; Wray, V.; Wagner, H. *Phytochemistry*, **1989**, *28*, 2373-2377.
13. Weisberger, A.S.; Pensky, J. *Science* **1957**, *126*, 1112-1114.
14. Siegers, C-P.; Steffen, B.; Röbke, A.; Pentz, R. *Phytomed.*, **1999**, *6*, 7-11.
15. Sendl, A.; Elbl, G.; Steinke, B.; Redl, K.; Breu, W.; Wagner, H. *Planta Med.* **1992**, *58*, 1-7.
16. Prasad, K.; Laxdal, V.A.; Yu. M.; Raney, B.L. *Molec. Cell. Biochem.* **1995**, *148*, 183-189.
17. Lee, C-H.; Parkin, K.L. *Food Chem.* **1998**, *61*, 345-350.
18. Lawson, L.D.; Wang, Z-Y.J.; Hughes, B.G. *Plant Med.* **1991**, *57*, 363-370.
19. Thomas, D.J.; Parkin, K.L. *Food Biotechnol.* **1991**, *5*, 139-159.
20. Synge, R.L.M.; Wood, J.C. *Biochem. J.* **1956**, *64*, 252-259.
21. Lancaster, J.E.; Kelly, K.E. *J. Sci. Food Agric.* **1983**, *34*, 1229-1235.
22. Carson, J.F.; Lundin, R.E.; Lukes, T.M. *J. Org. Chem.* **1966**, *31*, 1634-1635.
23. Shen, C.; Parkin, K.L. *J. Agric. Food Chem.* **2000** (in press).
24. Schwimmer, S.;Weston, W.J. *J. Agric. Food Chem.* **1961**, *9*, 301-304.
25. Schwimmer, S.; Ryan, C.A.; Wong, F.F. *J. Biol. Chem.* **1964**, *239*, 777-782.
26. Freeman, G.G.; Whenham, R.J. *J. Sci. Food Agric.* **1975**, *26*, 1333-1346.
27. Lawson, L.D. In *Garlic. The Science and Therapeutic Application of Allium sativum L. and Related Species,* 2nd ed.; Koch, H.P.; Lawson, L.D., Eds.; Williams & Wilkins: Baltimore, MD, 1996; pp 37-107.
28. Deleuze, H.; Langrand, G.; Millet, H.; Baratti, J. Buono, G.; Triantaphylides, C. *Biochim. Biophys. Acta.* **1987**, *911*,117-120.
29. Lancaster, J.E.; Shaw, M.L.; Randle, W.M. *J. Sci. Food Agric.* **1998**, *78*, 367-372.
30. Lawson, L.D.; Hughes, B.G. *Planta Med.* **1992**, *58*, 345-350.
31. Block, E.; O'Connor, J. *J. Am. Chem. Soc.* **1974**, *96*, 3929-3944.

Chapter 5

Interaction of Flavanols in Green Tea Extract during Heat Processing and Storage

Li-Fei Wang[1], Dong-Man Kim[1], and Chang Y. Lee[2]

[1]Korea Food Research Institute, Bundang Seongnam,
Kyonggi 463–420, Korea
[2]Department of Food Science and Technology, Cornell University,
Geneva, NY 14456

Epigallocatechin gallate (EGCG) is the most abundant and least stable flavanol in green tea extract. Understanding the chemical nature of EGCG and interrelationship with other flavanols during processing and storage could be important for the production of green tea beverage in cans and glass containers. EGCG added to the fresh green tea extract resulted in more EGC, ECG and EC after processing but much darker color than the control after heat processing and 12 days of storage. This could be due to the degradation of EGCG or the direct antioxidant effect of EGCG on the other epicatechins. Ascorbic acid added to the fresh green tea played some level of antioxidant function on flavanols, and decreased isomerization between catechins and epicatechins caused by heat treatment. However, owing to the extra color problems resulted from the oxidation of ascorbic acid itself and its reactions with other green tea components, application of ascorbic acid in the production of green tea beverage in cans or glass containers still requires further research

Green tea flavanols have recently received much attention due to their pharmaceutical functions, such as antioxidative, antitumor, and anticarcinogenic activities (*1-7*). Recently, food companies are packaging green tea extract in cans or glass bottles in order to increase the consumption of green tea. However, it was noted that the canned green tea extracts tended to lose their sensory qualities, mainly due to the instability or polyphenolics during heat processing and storage (*8*)

Up to 90% of the green tea polyphenolics is composed of flavanols (*9*). (-)-Epigallocatechin gallate (EGCG), (-)-Epigallocatechin (EGC), (-)-epicatechin gallate (ECG), (-)-epicatechin (EC), (+)-catechin (C) and (+)-gallocatechin (GC) are the major flavanols mostly discussed previously (*9-11*). Seto et al. (*12*) reported that green tea flavanols undergo isomerization between catechins and epicatechins by high temperature treatment at 120°C. This isomerization involves a change in configuration at the 2-postion without changing the optical rotation (Figure 1). The similar reaction can be expected between epicatechin (EC) and catechin, epicatechin gallate (ECG) and catechin gallate (CG), epigallocatechin (EGC) and gallocatechin (GC), and epigallocatechin gallate (EGCG) and gallocatechin gallate (GCG).

In our previous study (*8*), it was found that the amounts of catechins (including catechin, CG, GC and GCG) in green tea extracts were initially increased while epicatechins (EC, ECG, EGC and EGCG) were decreased by heat processing (at 121°C for 1 minute). This is due to the isomerization of epicatechins to catechins. However, all flavanols in processed tea extracts were gradually decreased during storage (50°C for up to 12 days). The severity of degradation of epicatechins was in the order of EGCG>EGC>ECG>EC. Since EGCG is the major flavonoid in green tea and it has a great effect on the sensory qualities of green tea extract, the objectives of this research were to investigate the effects of processing and storage on the chemical nature of EGCG and its interrelationship with other flavanols in green tea extract.

Experimental

The Control Samples

Harvested fresh tea leaves (*Camellia sinesis*) were immediately processed into commercial steamed green tea in August, 1998 at Posung, Korea. Green tea extracts were prepared from green tea (tea leaves:distilled water = 1:160,

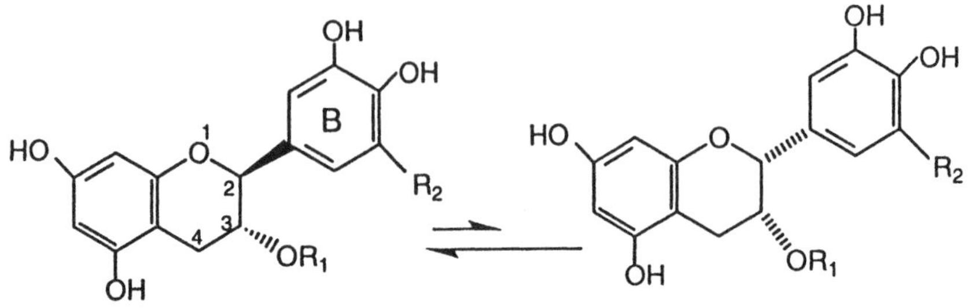

Catechins (2,3-trans)   Epicatechins (2,3-cis)

(-)-Catechin (C): $R_1 = R_2 = H$ (-)-Epicatechin (EC): $R_1 = R_2 = H$

(-)-Catechin gallate (CG):   (-)-Epicatechin gallate (ECG):
$R_1 = X, R_2 = H$ $R_1 = X, R_2 = H$

(-)-Gallocatechin (GC):   (-)-Epigallocatechin (EGC):
$R_1 = H, R_2 = OH$ $R_1 = H, R_2 = OH$

(-)-Gallocatechin gallate (GCG):   (-)-Epigallocatechin gallate (EGCG):
$R_1 = X, R_2 = OH$ $R_1 = X, R_2 = OH$

$$X = \text{gallic acid} = -\overset{\displaystyle O}{\overset{\|}{C}} - \text{(OH, OH, OH)}$$

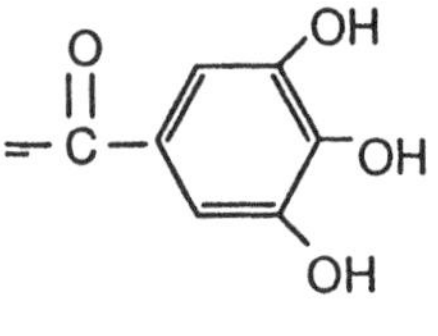

Figure 1. Isomerization of flavanols caused by heat treatments. Note: The original catechins existing in fresh tea leaves have the (+)-optical rotation property, while epicatechins are in the (-)-form. After heat treatments, (-)-catechins and (+)-epicatechins could be formed from the isomerization of (-)-epicatechins and (+)-catechins, respectively.

w/w) by extracting at 80°C for 4 minutes (8), and the filtrates (fresh extracts) were further processed:

FT: fresh green tea extract

PT: fresh tea extract was heat processed (at 121°C/1 min)

PT6 and PT12: heat processed tea extracts stored at 50oC for 6 and 12 days. Respectively.

Influences of Additional EGCG on the Green Tea Extracts

EGCG is the most important flavanol in deciding the sensory qualities of green tea extracts (8). In our previous experiment, it was found that about 0.65 mg EGCG/mL tea extract was lost during processing and storage. To observe how this amount of EGCG influences the quality of green tea extracts, 0.65 mg EGCG/mL was added to the FT extract and then heat processed and stored.

Effect of Ascorbic Acid on the Processed Green Tea Extracts

Since ascorbic acid is the common antioxidant applied in food processing, it was added to FT at the level of 0.4% and then the tea extract was heat processed and stored. Samples were taken periodically, sealed in plastic bags and kept in a −70°C freezer until analyzed.

Analysis of Flavanol Compounds

Tea samples were pretreated and analyzed for flavanols by HPLC according to the methods of Matsuzaki et al. (4), Ikegaya et al. (13) and Wang et al. (8). Identification and quantification of flavanols were achieved by comparing retention times, spectra and peak area of the chromatograms with the authentic references.

Measurement of Color

The color of green tea extracts was measured using the HunterLab color meter (ColorQuest II, Hunter Associates Laboratory, Inc., Reston, VA), and expressed as L, a, b values.

62

Results and Discussion

Interrelationship of EGCTG with Other Flavanols in Green Tea Extracts

As shown in Figure 2, the addition of EGCG to fresh tea extract significantly increased the GCG content during heat processing. This increase was about 2.5 times higher than that of the control. This confirms our previous conclusion that the increase of catechins in green tea extracts by high temperature treatments was resulted from the isomerization of epicatechins (8). During storage, GCG in both control and EGCG-added samples decreased in the same rate so that GCG in the latter sample still remained more than two folds of the former one after 12 days of storage.

The addition of EGCG also has a direct effect on the concentrations of other epicatechins. More EGC, ECG and EC were detected in the sample added with EGCG during processing and storage (data not shown). This may be due to in part of heat degradation of EGCG, hydrolysis of EGCG may result in EGC and gallic acid (Figure 3). On the other hand, some flavanols are known as good antioxidants and EGCG has shown to have the lowest redox potential among the green tea epicatechins (14). Therefore, EGCG appeared to function as an antioxidant for the other flavanols when mixed together in the aqueous media. The additional EGC, ECG and EC detected in the sample treated with EGCG could be resulted from protection by EGCG and/or oxidation of EGCG. The oxidation rate of EGCG was found to be concentration dependent; at a higher starting concentration, EGCG oxidized faster (data not shown). In addition, the oxidation of EGCG had a profound effect on color. When 0.65 mg EGCG/mL was added, it significantly darkened the color of green tea extract at the final stage of storage (PT12), and the oxidation products of EGCG appeared to contribute more to red color than to yellow in the green tea extract.

Figure 4 shows the loss (%) of various epicatechins, based on their original amounts in FT, during processing and storage. EGCG and EGC are very unstable, so that even without heat treatment when FT was stored at 50°C for 12 days, they lost 87% and 73%, respectively. Losses of ECG and EC in FT stored at the same conditions were only 36% and 15%, respectively. Heat processing accelerated the destruction of epicatechins. With processing at 121°C for 1 minute and 12-day storage, the total losses of EGCG and EGC were up to 96% and 89%, respectively, in which 52% of EGCG and 48% of EGC were damaged by heat treatment. Compared to EGCG and EGC, ECG and EC were more stable to heat processing with total loss of 67% and 53%, respectively. On the other hand, concentrations of catechins (GC, C, GCG and

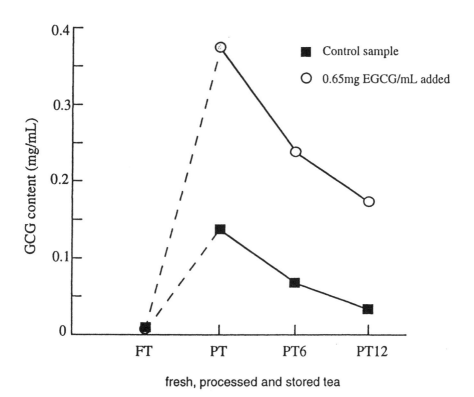

Figure 2. Effect of EGCG addition on the changes of GCG in green tea extract during processing and storage. (-------: Changes caused by heat processing; ———: changes caused by storage).

64

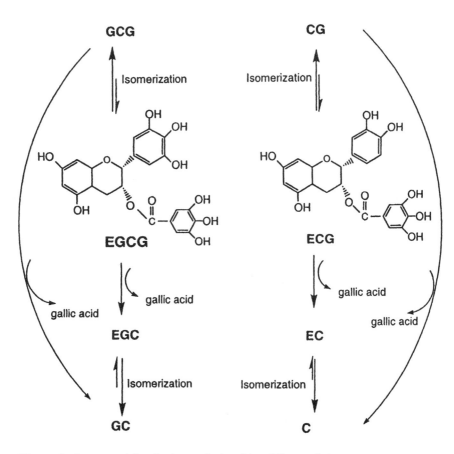

Figure 3. A proposal for the interrelationship of flavanols in green tea extracts during processing and storage.

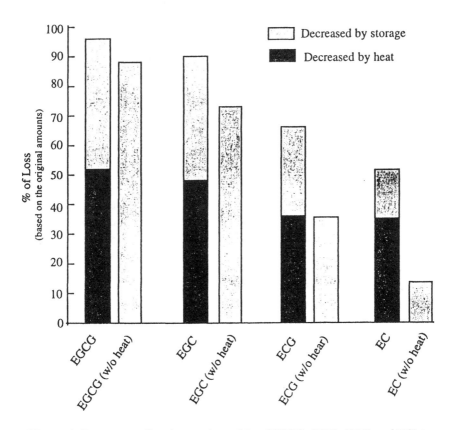

Figure 4. Decreases of various epicatechins (EGCG, EGC, ECG and EC) in the green tea extracts with and without heat processing after 12 days of storage.

CG) were increased due to the isomerization of epicatechins by high heat treatment, as mentioned previously. Without heat processing, these catechins could be also increased slightly during storage, even though the quantity of increase was much less and the formation rate was much slower. Isomerization of flavanols has been reported to take place minimally at the temperature of 80°C or below (*12*). Our results indicated that flavanol isomerization may occur at temperatures as low as 50°C, if the storage time is long enough.

Effect of Ascorbic Acid on Green Tea Extracts

Ascorbic acid has been used commercially as an antioxidant in canned green tea beverage. The addition of 0.4% ascorbic acid decreased the pH of FT from 6.01 to 3.25. No significant change of pH was found after processing, however it was further reduced to 2.82 after 12 days of storage. It has been revealed that green tea flavanols could be kept very stable in an acidic solution (pH<4), whereas the alkaline media (pH>8) has a negative impact on their preservation (*15,16*). It is believed that the acidity of ascorbic acid also plays an important role in its antioxidative effect on the green tea flavanols. As shown in Figures 5 and 6, ascorbic acid can effectively preserve epicatechins from heat losses. Zhu et al. (*16*) reported that the effect of ascorbic acid was dose-dependent, and the addition of ascorbic acid did not regenerate green tea flavanols, but only reduced the rate of flavanol degradation. However, some epicatechins (EGC, EC and EGCG) in our ascorbic acid-added samples slightly increased after heat treatments. It must be noted that Zhu et al. did not treat the samples at high temperature. Instead, they incubated the partially purified green tea flavanols with ascorbic acid at 37°C. Ascorbic acid also appeared to prohibit flavanol isomerization caused by heating. With the addition of 0.4% ascorbic acid, GC increase was less than the control after heat processing and it also decreased more slowly during storage. The production of ascorbic acid on other catechins was more effective, so that C, GCG and CG did not change significantly during the whole observation period. Even though ascorbic acid showed the good antioxidative effect on green tea flavanols, it was not an effective antibrowning reagent for the processed green tea beverage because addition of ascorbic acid further changed the color of green tea extracts. As soon as ascorbic acid (0.4%) was added to the fresh green tea extracts, a light pink color was visually observed. After heat processing, the ascorbic acid-added sample had a color that was lighter, less red and yellow than the control, but lacked the attractive greenish color of fresh green tea extracts. The variation in colors of fresh tea and heat processed tea might be resulted from the interactions between the green tea components and oxidation products of ascorbic acid, because their corresponding reference samples, the non-treated and processed 0.4% ascorbic acid aqueous solutions, remained colorless. During storage, the color of samples added with ascorbic acid gradually turned brown, but showed less dark, red and more yellow than the control sample at the final stage of storage. During the 12-day storage period, the colored oxidation products from ascorbic acid gradually appeared and also attributed to the color variation of ascorbic acid-added samples.

Conclusion

EGCG is the dominant factor among the flavanols that affects composition of other flavanols and the sensory qualities of processed green tea extract. Therefore, the content of EGCG in fresh green tea extract can be used as a processing index for the qualities of processed green tea extracts. During the processing of green tea beverage, the heat treatments significantly affected the chemistry of flavanols by destroying more than 50% of EGCG in one minute when heated at 121°C. Ascorbic acid showed some level of antioxidant effect on flavanols (especially epicatechins), but its application on the green tea beverage is still remained problematic because the oxidation of ascorbic acid itself and its reactions with other green tea components could cause extra color changes.

References

1. Chung, F.; Xu, Y.; Ho, C.-T.; Desai, D.; Han, C. In *Phenolic Compounds in Food and Their Effects on Health II*, Huang, M.T.; Ho, C.-T.; Lee, C.Y., Eds.; ACS Symp. Ser. No. 507, American Chemical Society: Washington, D.C., 1992, pp. 300-307.
2. Chung, K.; Wei, C.; Johnson, M.C. *Food Sci. Technol.* **1998**, *9*, 168-175.
3. Conney, A.H.; Wang, Z.Y.; Ho, C.-T.; Yang, C.S. Huang, M.T. In *Phenolic Compounds in Food and Their Effects on Health II*, Huang, M.T.; Ho, C.-T.; Lee, C.Y., Eds.; ACS Symp. Ser. No. 507, American Chemical Society: Washington, D.C., 1992, pp. 284-291.
4. Matsuzaki, T.; Hara, Y. *Nippon Nogeikagaky Kaishi.* **1985**, *59(2)*, 129-134.
5. Prochaska, H,J,; Talalay, P. In *Phenolic Compounds in Food and Their Effects on Health II*, Huang, M.T.; Ho, C.-T.; Lee, C.Y., Eds.; ACS Symp. Ser. No. 507, American Chemical Society: Washington, D.C., 1992, pp. 150-159.

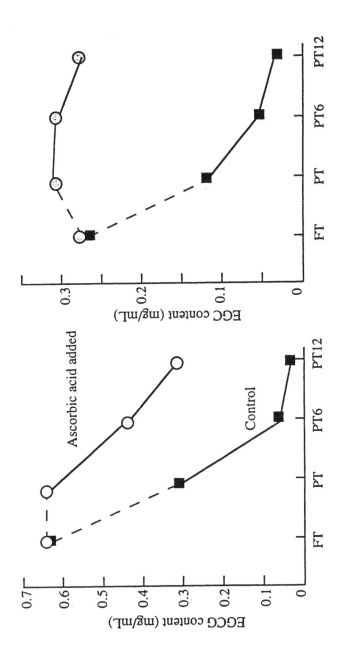

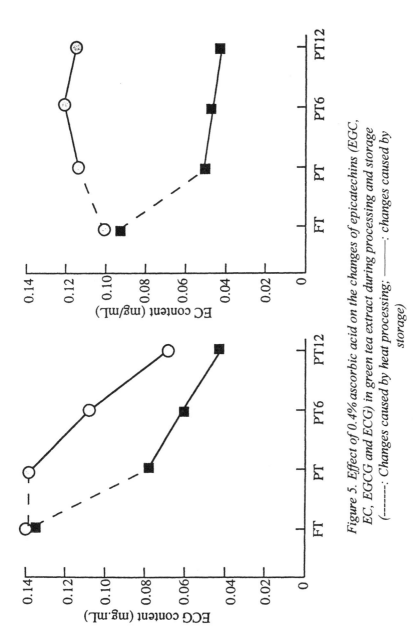

Figure 5. Effect of 0.4% ascorbic acid on the changes of epicatechins (EGC, EC, EGCG and ECG) in green tea extract during processing and storage (------: Changes caused by heat processing; ——: changes caused by storage)

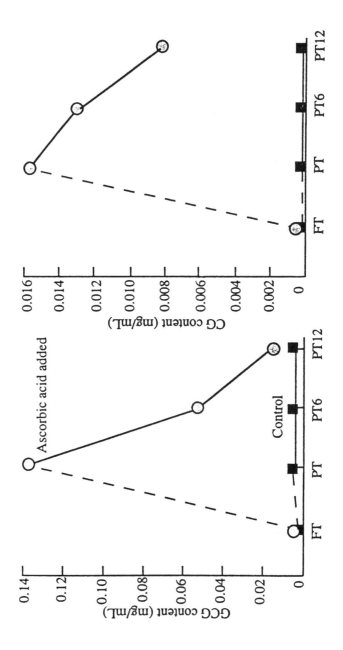

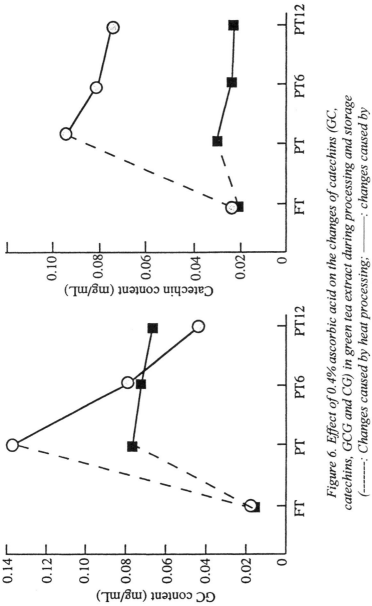

Figure 6. Effect of 0.4% ascorbic acid on the changes of catechins (GC, catechins, GCG and CG) in green tea extract during processing and storage (------: Changes caused by heat processing; ——: changes caused by storage)

6. Sakanaka, S.; Kim, M.; Taniguchi, M.; Yamamoto, T. *Agric. Biol. Chem.* **1989**, *53*, 2307-2311.

7. Wang, Z.Y.; Hong, J.; Huang, M.T.; Conney, A.H.; Yang, C.S. In *Phenolic Compounds in Food and Their Effects on Health II*, Huang, M.T.; Ho, C.-T.; Lee, C.Y., Eds.; ACS Symp. Ser. No. 507, American Chemical Society: Washington, D.C., 1992, pp. 292-299.

8. Wang, L.; Kim, D.; Lee, C.Y. *J. Agric. Food Chem.* Submitted.

9. Yamanishi, T.; Hara, Y.; Luo, S.; Wickremasinghe, R.L. *Food Rev. Internat.* **1995** *11*, 371-546.

10. Balentine, D.A. In *Phenolic Compounds in Food and Their Effects on Health I*, Ho, C.-T.; Lee, C.Y.; Huang, M.T. Eds.; ACS Symp. Ser. No. 506, American Chemical Society: Washington, D.C., 1992, pp. 102-117.

11. Lunder, T.L. In *Phenolic Compounds in Food and Their Effects on Health II*, Huang, M.T.; Ho, C.-T.; Lee, C.Y., Eds.; ACS Symp. Ser. No. 507, American Chemical Society: Washington, D.C., 1992, pp. 114-120.

12. Seto, R.; Nakamura, H.; Nanjo, F.; Hara, Y. *Biosci. Biotech. Biochem.* **1997**, *61*, 1434-1439.

13. Ikegaya, K.; Takayanagi, H.; Anan, T. *Bull. Natl. Res. Tea* **1990**, *71*, 43-74.

14. Opie, S.C.; Clifford, M.N.; Robertson, A. *J. Sci. Food Agric.* **1993**, *63*, 435-438.

15. Suematsu, S.; Hisanobu, Y.; Saigo, H.; Matsuda, R.; Hara, K. *Nippon Shokuhin Kogyo Gakkaishi.* **1993**, *63*, 435-438.

16. Zhu, Q.Y.; Zhang, A.; Tsang, D.; Huang, Y.; Chen, Z.Y. *J. Agric. Food Chem.* **1997**, *45*, 4624-4628.

Chapter 6

Influence of Processing Conditions on Isoflavones Content of Soybean Products

Gow-Chin Yen and Chia-Horn Kao

Department of Food Science, National Chung Hsing University, 250 Kuokuang Road, Taichung 402, Taiwan

The effects of processing conditions on the changes in isoflavones content of black soymilk and soymilk film (yuba) were investigated. When black soybeans were soaked in water at 30 and 50 °C for various periods of time, the contents of daidzein and genistein increased with an increase of the soaking time while daidzin and genistin decreased. There was no significant difference ($P < 0.05$) in the contents of isoflavones in black soybean under soaking at 30 °C for 12 h and at 50 °C for 6 h. The amounts of isoflavones in black soybean changed markedly under soaking at 20-60 °C for 8 h. The change in the contents of isoflavones in black soybean during soaking was related to its β-glucosidases activity. The effect of soaking temperature in β-glucosidases activity of black soybean was in the order of 50 > 40 > 60 > 30 > 20 °C. The contents of daidzein and genistein in Yuba prepared by soaking soybeans at 50 °C for 6 h were about 2.5 times of that prepared by soaking soybeans at 30 °C for 12 h. Thus, changing the processing conditions can increase the contents of isoflavone aglycones in soybean products.

Soybeans have many unique phytochemicals, including isoflavones, saponins, phytates, phytosterols, phenolic acids, and trypsin inhibitors. All of these compounds have been mentioned as having important beneficial effects in the prevention of degenerative conditions, such as heart disease and cancer (*1*). The isoflavones in soybean, primarily genistein and daidzein, have been well researched by scientists for their antioxidant and phytoestrogenic properties (*2*). A large number of epidemiological studies have also shown that isoflavones may reduce the risk of hormone-dependent cancers, such as breast and prostate cancer, as well as other cancers (*3-5*). Therefore, the isoflavones may come to be regarded as phytochemicals important for human health in the future.

The contents of isoflavones in soybean products have been reported by many researchers (*6-8*). The data indicate that the levels of genistein and daidzein in some fermented soy foods, soymilk, and tofu are higher than that in soybean. Most reports have also indicated that the higher levels of genistein and daidzein were in soybean products, the lower were the levels of genistin and daidzin. These results might be due to the fact that genistin and daidzin (glucosides) were hydrolyzed to genistin and daidzin (aglycon) by β-glucosidase (*9,10*). The activity of β-glucosidase in soybean was found to be influenced by processing and fermentation, which resulted in changes in isoflavones content. Ikeda et al. (*11*) studied the changes in isoflavones during miso fermentation. Wang and Murphy (*12*) reported on the effects of processing techniques on the distribution of isoflavones in manufacturing temph, soymilk, tofu and protein isolate. However, data are not available concerning the effects of modifications of existing processing techniques on the retention or increase of isoflavones of soybean products during processing. In recent years, black soymilk (made from black soybean) is a very popular commercial product in Taiwan. In addition, the soymilk film (Yuba) is a traditional soyproduct that consumed in Asian countries. Therefore, the objective of this study was to investigate the influence of processing conditions, especially the soaking conditions, on changes in isoflavones during the manufacturing of black soymilk and soymilk film.

Experimental

Materials and Chemicals. Soybean and black soybean were purchased from a local market in Taichung, Taiwan. Genistin, Genistein, Daidzein, 6''-*O*-malonyldaidzin and 6''-*O*-malonylgenistin, fluorescein, β-nitrophenol-β-*D*-glucopyranoside, and *p*-nitrophenol were purchased from the Sigma Chemical Company (St. Louis, MO). Daidzin was obtained from the Extrasynthese Company (Geany Cedex, France).

Effect of Soaking Conditions. Black soybeans (25 g) were soaked in 100 mL of distilled water for 0, 4, 8, 12, and 16 h, respectively, at 30 °C. To study the effect of the soaking temperature on changes of isoflavones in black soybean, black soybeans were soaked at 20, 30, 40, 50, and 60 ℃, respectively, for 8 h. To study the effect of the soaking time on changes of isoflavones in black soybean, black soybeans were soaked at an optimum temperature of 50 ℃ for 0, 2, 4, 6, and 8 h, respectively. Each soaked sample was cooled and ground in a Waring blendor. The slurry was freeze-dried for the following isoflavones analysis. Each treatment was duplicated.

Preparation of Black Soymilk. Black soybeans (250 g) were soaked in 1000 ml of distilled water at 30 ℃ for 12 h or soaked at 50 ℃ for 6 h. The soaked soybeans were rinsed, drained and ground with water. The slurry was cooked and filtered to separate the soymilk from the water-insoluble residue. The soymilk was used for sensory evaluation or freeze-dried for isoflavones analysis.

Preparation of Soymilk Film (Yuba). Washed soybeans (200 g) were soaked in 800 mL water at 30 ℃ for 12 h or at 50 ℃ for 6 h. The soaked soybeans were blended with an appropriate volume of water and then cooked; through filtration, the soymilk and residue were separated. When the soymilk was heated to over 80 ℃, a thin film formed on the surface of the soymilk; this film is called soymilk film (yuba). The soymilk film could be picked up with a piece of bamboo at 10 min intervals. During the process of making soymilk film, we sampled at different steps of after soaking, filtered soy residue and soymilk, soymilk film, and residual soymilk. The samples were blended and freeze-dried; then the isoflavones content were determined by HPLC.

Isoflavones Extraction. Isoflavones were extracted and analyzed using the method of Coward et al. (7) with a modification. A freeze-dried sample (1 g) was mixed with 10 mL of 80% methanol and 1.5 mg of fluorescein (as an internal standard), stirred at 60 ℃ for 1 h, and then centrifuged at 3800 × g for 10 min. The upper solution was dried under vacuum. The dried materials was redissolved in 5 mL of 50% methanol and extracted with 20 mL hexane four times to remove lipids. The methanol layer was evaporated to dryness and then redissolved in 10 mL of 80% methanol. The solution was filtered through a 0.45 μm filter paper before HPLC analysis.

HPLC Analysis. The HPLC system consisted of a Hitachi model L-6200 intelligent pump, a Rheodyne model 7125 injector, a Hitachi model L-4200 UV-vis detector, and a Hitachi model D-2500 Chromato-integrator. The column was a LiChrosorb 100 RP-18 (5 μm, 250 × 4.6 mm i.d., E. Merck). The solvent system used was a gradient of 15-45% acetonitrile in 0.1% (v/v) trifluoroacetic acid at a flow rate of 1.3 mL/min. The concentration of acetonitrile increased at a rate of 1%/min. The detecting wavelength was set at 262 nm. The concentrations of isoflavones were calculated from standard curves of the area responses for authentic isoflavone standards normalized to the constant amount of fluorescein added to each sample.

Determination of β-Glucosidase Activity. The activity of β-glucosidase was measured according to the method of Heyworth and Walker (*13*). Whole black soybeans (10 g) in a beaker were soaked with 30 mL distilled water and 10-mg β-nitrophenol-β-*D*-glucopyranoside (β-NPG) at different temperatures (20-60 ℃) for 8 h. The soaking solution was decanted and the soybeans were washed with deionized water. The soybeans were then cooked in 100 mL boiling water to inactivate the enzyme activity. After cooling and dehulling, the soybeans were soaked in 80 ml 0.25 M sodium carbonate for 2 h to extract the *p*-nitrophenol that was hydrolyzed from β-NPG by β-glucosidases in the soybeans. The resulting yellow color was measured at 420 nm with a spectrophotometer (Hitachi U-2000, Japan). The amount of hydrolyzed *p*-nitrophenol was determined by referring to a standard curve. A control group without the addition of β-NPG was conducted for each temperature treatment.

Sensory Evaluation of Black Soymilk. The sensory quality of the black soymilk prepared under the two different soaking conditions (30 ℃, 12 h and 50 ℃, 6 h) was evaluated by means of the Preference test and Paired comparison test. Each sample was added with 7% sugar and kept at 6-10 ℃. The preference test was evaluated using a 7 point scale and 30 panelists. The paired comparison test was conducted using 15 panelists and 10-20 sec intervals for each sample.

Statistical Analysis. All the sample analyses were run in triplicate. Data were analyzed using the Statistical Analysis System software package. Analysis of variance was performed using ANOVA procedures. Significant differences ($P < 0.05$) between means were determined using Duncan's multiple range test.

Results and Discussion

Black Soybean Soaked at 30 ℃ with Different Time Periods

Because room temperature in summer in Taiwan is around 30 ℃, a temperature of 30 ℃ was selected to study the effect of different soaking times on the contents of isoflavones in black soybean. The contents of isoflavones in black soybean soaked at 30 ℃ for different time periods is shown in Figure **1** and expressed in mg/g dry wt. The results indicate that the aglycon of isoflavones, daidzein and genistein, in black soybean increased with increasing soaking time, while the glycoside of isoflavones, daidzin and genistin, decreased. The contents of daidzein and genistein in black soybean soaked at 30 °C for 4 h

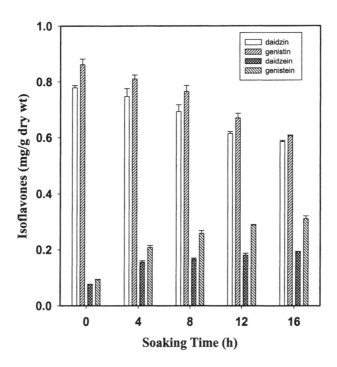

Figure 1. Changes in isoflavones of black soybean during soaking perios at 30 °C.

increased significantly ($P < 0.05$) from 76 and 92 to 156 and 208 µg/g dry wt, respectively, when compared with the unsoaked sample. After soaking for 16 h, the contents of daidzin, genistin, daidzein, and genistein in black soybean were 585, 606, 193, and 311 µg/g dry wt, respectively. Matsura et al. (9) also reported that the contents of daidzein and genistein in soybean increased with an increase of soaking time at a temperature of 20 °C. In addition, the total isoflavones in black soybean decreased slightly with increasing soaking time; the total isoflavones in black soybean before and after soaking for 16 h were 1808 and 1695 µg/g dry wt, respectively.

78

Black Soybean Soaked at Different Temperatures

The results shown in Figure 2 indicated that the contents of daidzein and genistein in black soybean increased with the soaking temperature from 20 to 50 °C and then decreased at 60 °C. The contents of daidzein and genistein in black soybean increased about 2.7-fold when the soaking temperature increased

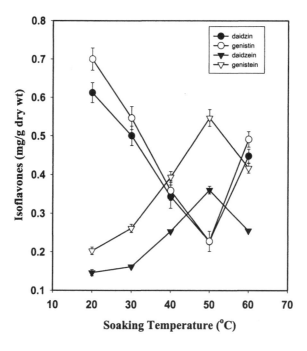

Figure 2. Effect of soaking temperature in isoflavones of black soybean during the soaking (soaking time 8 h).

from 20 to 50 °C. In contrast, the contents of daidzin and genistin in black soybean decreased with the soaking temperature from 20 to 50 °C and then increased at 60 °C. The total isoflavones in black soybean soaked at 20, 30, 40, 50, and 60 °C were 1661, 1469, 1348, 1361, and 1610 µg/g dry wt, respectively. The fluctuation in total isoflavones in black soybean with different soaking temperatures might have been due to changes in the activity of β-glucosidases. Since soybean contains β-glucosidase, which can hydrolysis daidzin and genistin to daidzein and genistein (*9,10*), the effect of the soaking temperature on the activity of β-glucosidase in black soybean should be evaluated to reveal its effect on total isoflavones.

Activity of β-Glucosidases in Black Soybean

β-Glucosidase is able to hydrolysis p-nitrophenyl-β-D-glucopyranoside to form p-nitrophenol. p-Nitrophenol appears yellow color in alkaline solution and can be measured with absorbance at 420 nm. The activity of β-glucosidase can be determined from the standard curve of p-nitrophenol. The linear regression of standard curve of p-nitrophenol is is $Y=1.5564\times10^{-2}X-7.5382\times10^{-3}$, and the correlation efficient is $r=0.9994$. Figure **3** shows the effect of the soaking

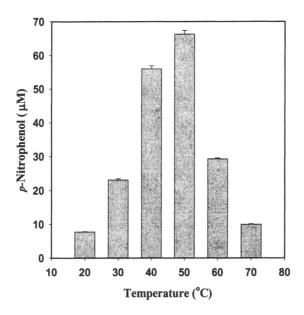

Figure 3. Effect of temperature on the action of β-glucosidase to p-nitrophenyl-β-D-glucopyranoside during soaking of black soybeans.

temperature on the activity of β-glucosidase in black soybean to hydrolysis of p-nitrophenyl-β-D-glucopyranoside. The results indicate the concentration of p-nitrophenol increased with increasing the soaking temperature up to 50 °C and then decreased at 60 °C. This result means that the black soybean had the highest level of β-glucosidase activity when the soaking temperature was 50 °C.

80

Matsurra and Obata (*10*) demonstrated that β-Glucosidase hydrolized the daidzin and genistin to daidzein and genistein in soymilk, and that its effect on genistin was greater than that on daiazin. Therefore, the activity of β-glucosidase in black soybean during the soaking process is the main factor influencing its contents of isoflavones.

Black Soybean Soaked at 50 ℃ for Different Time Periods

The effect of the soaking time on the total isoflavones in black soybean soaked at 50 °C is shown in Figure **4**. As the soaking time increased, the contents of daidaein and genistein increased while the contents of daidzin and genistin decreased. The contents of daidzin, genistin, daidaein, and genistein in

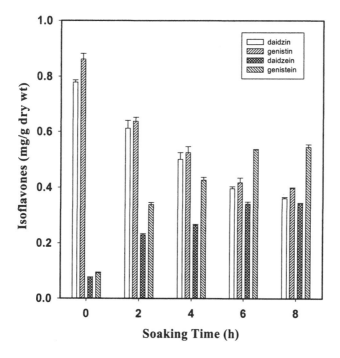

Figure 4. Changes in isoflavones of black soybean during saoking period at 50 ℃.

black soybean soaked at 50 °C for 2 h were 611, 637, 223, and 339 µg/g dry wt, respectively, whereas they were 361, 398, 344, and 545 µg/g dry wt under soaking for 8 h. Comparing the results shown in Fig. 2, it can be seen that there was no significant ($P > 0.05$) difference in the total isoflavones in black soybean under soaking at 50 °C for 2 h and at 30 °C for 12 h. This means that the effect of the soaking temperature on the total isoflavones in black soybean was greater than that of the soaking time. In addition, the total isoflavones in the black soybean decreased with increasing the soaking time: 1817, 1717, 1691, and 1648 μg/g dry wt of isoflavones under soaking at 50 °C for 2, 4, 6 and 8 h, respectively.

Isoflavones Content and Sensory Evaluation of Black Soymilk

The black soybean soaked at 30 °C for 12 and at 50 °C for 6 h was processed into soymilk and named as black soymilk A and B, respectively. There was a significant ($P < 0.05$) difference in total isoflavones between these two products (Table I). The contents of daidzein and genistein in black soymilk B was 1.9-fold higher than that in black soymilk A, whereas the contents of daidzin and genistin in black soymilk B was only 60% of that in black soymilk A. Many researchers have studied the off -flavor of soybean. Arai et al. (*14*) isolated some phenolic acids from soymeal, which were related to the sour, bitter and astringency taste of soybean. Sessa et al. (*15*) indicated that the oxidation of phosphatidylcholines was the major factor related to the bitter taste of soybean. Huang et al. (*16*) identified daidzein, genistein and glycitein-7-β-glucoside from defatted soymeal, and they also recognized that these compounds caused the bitter and astringency taste of soybean products. As the results discussed above show, the amounts of the bioactive compounds, daidzein and genistein, in black soymilk can be increased by changing the processing conditions; however, the sensory acceptance of products also should be evaluated. In this study, sensory analysis of black soymilk A and B was conducted using the Preference test and Paired compairson test. The results

Table I. The Isoflavones Content of Black Soymilk Prepared with Different Processes

Sample [a]	Isoflavones (μg/g dry wt) [b]			
	Daidzin	Genistin	Daidzein	Genistein
Soymilk A	614±28	670± 20	180± 4	287± 7
Soymilk B	396±24	418± 17	341± 6	535±19

[a] Soymilk A: Soymilk was prepared from black soybeans which soaked at 30 °C for 12 h.

Soymilk B: Soymilk was prepared from black soybeans which soaked at 50 °C for 6 h.

[b] Values are means± SD of three replicate analyses.

indicated that there was no significant ($P > 0.05$) difference in the sensory evaluation scores between these two products.

Changes in the Contents of Isoflavones during the Processing of Yuba

The Yuba was manufactured with two different processing conditions. The difference between these two processes was in the soaking conditions, one was soaking at 30 °C for 12 h and the other was soaking at 50 °C for 6 h. During the process, the total isoflavones in the soybeans after soaking, the soymilk and filtrate, Yuba, and residual milk were analyzed. Table II shows the changes in isoflavones in Yuba during processing. The contents of daidzin, genistin, daidzein, and genistein in soybean after soaking at 30 °C for 12 h were 679, 605, 112, and 155 µg/g dry wt., respectively, while they were 391, 313, 306, and 481 µg/g dry wt after soaking at 50 °C for 6 h. The Yuba made under these two soaking conditions was named Yuba A (30 °C for 12 h) and Yuba B (50 °C for 6 h). The contents of daidzein and genistein in Yuba B were 389 and 500 µg/g dry wt, respectively, which was 2-fold greater than that in Yuba A. In addition, during slow heating of soymilk to make Yuba, the contents of daidzein and genistein in residual soymilk were reduced.

During the Yuba manufacturing process, the soymilk should be continually heated (> 85 °C) to make it possible to successively pick up the Yuba. In this

Table II. The Contents of Isoflavones in Yuba during Processing

Step	Isoflavones (µg /g dry wt) [a]							
	Daidzin		Genistin		Daidzein		Genistein	
	A[b]	B[b]	A	B	A	B	A	B
Soaked soybeans	679	391	605	313	112	306	155	481
Residue	409	221	414	243	75	234	120	356
Soymilk	779	567	685	495	99	291	143	407
Yuba	922	845	1026	956	141	389	218	505
Residual soymilk	1251	1114	1127	1001	93	214	141	186

[a] A: Soybeans were soaked at 30 °C for 12 h to prepare yuba. B: Soybeans were soaked at 50 °C for 6 h to prepare yuba.

[b] Values are means of three replicate analyses.

experiment, it was usually possible to pick up 15 pieces of Yuba during continual heating for 2-2.5 h. However, there was a great change in the contents of daidzin and genistin during the heating process. As the results shown in Table II, the contents of daidzin and genistin in Yuba and residual soymilk were obviously higher than those in soymilk, indicating that some compounds in soymilk may be degraded into daidzin and genistin during long periods of heating. Recently, Barnes et al. (*17*) indicated that soybean cooked with hot water caused 6''-*O*-malonylgenistin to hydrolyze to genistin during the soymilk manufacturing process. Therefore, we analyzed the changes in 6''-*O*-malonyldaidzin and 6''-*O*-malonylgenistin in each step of the Yuba manufacturing process, and the results are shown in Table **III**. The contents of 6''-*O*-malonyldaidzin and 6''-*O*-malonylgenistin in Yuba and residual soymilk were lower than those in soymilk; however, the contents of daidzin and genistin increased. Thus, during continual cooking of soymilk to make Yuba, the increases of daidzin and genistin were due to the hydrolysis of 6''-*O*-malonyldaidzin and 6''-*O*-malonylgenistin, which removed the malonate ester group. In our previous study, we also found that commercial Yuba contained the highest amounts of daidzin and genistin among the soybean products tested (*18*).

Table III. The Contents of Malonyldaidzin and Malonylgenistin in Yuba during Processing

Step	Isoflavones ($\mu g/g$ dry wt) [a]			
	Malonyldaidzin		Malonylgenistin	
	A^b	B^b	A	B
Soaked soybeans	627	565	704	639
Residue	263	184	273	214
Soymilk	665	720	722	682
Yuba	183	23	227	44
Residual soymilk	388	100	430	158

a, b The descriptions are the same as in Table II.

Conclusion

The activity of β-glucosidase in black soybean during the soaking process is the main factor influcing its contents of isoflavones. The effect of the soaking temperature on the total isoflavones in black soybean was greater than that of the soaking time. The contents of daidzein and genistein in Yuba were found to be related to their contents in soybeans after soaking. The increases of daidzin and genistin in Yuba and residual soymilk were due to the hydrolysis of 6''-*O*-malonyldaidzin and 6''-*O*-malonylgenistin, respectively, with long periods of heating. The amounts of the active compounds, daidzein and genistein, in black soymilk and soymilk film (Yuba) can be increased by changing the processing conditions.

84

References

1. Wang, C.; Wixon, R. *INFORM.* **1999**, *10*, 315-321.
2. Wei, H.; Bowen, R.; Cai, Q.; Barnes, S.; Wang, Y. *Proc. Soc. Exp. Biopl Med.* **1995**, *208*, 124-129.
3. Adlercreutz, H., Honjo, H., Higashi, A., Fotsis, T., Hmlinen, E., Hasegawa, T.; Okada, H. *Am. J. Clin. Nutr.* **1991**, *54*, 1093-1100.
4. Lee, H. P.; Gourlery, L.; Duffy, S. W.; Esteve, J.; Day, N. E. *Lancet* **1991**, *337*, 1197-1200.
5. Messina, M. J.; Persky, V.; Setchell, K. D.; Branes, S. *Nutri. Cancer* **1994**, *21*, 113-131.
6. Murphy, P. A. *Food Technol.* **1982**, *36(1)*, 60, 62-64
7. Coward, L.; Barnes, N. C.; Setchell, K. D. R.; Barnes, S. *J. Agric. Food Chem.* **1993**, *41*, 1961-1967.
8. Fukutake, M.; Takahashi, M.; Ishida, K.; Kawamura, H.; Sugimura, T.; Wakabayashi, K. *Fd. Chem. Toxic.* **1996**, *34*, 457-461.
9. Matsuura, M.; Obata, A.; Fukushima, D. *J. Food Sci.* **1989**, *54*, 602-605.
10. Matsuura, M.; Obata, A. *J. Food Sci.* **1993**, *50*, 144-147.
11. Ikeda, R.; Ohta, N.; Watanabe, T. *Nippon Shokuhin Kagaku Kogaku Kaishi* **1995**, *42*, 322-327.
12. Wang, H. J.; Murphy, P. A. *J. Agric. Food Chem.* **1996**, *44*, 2377-2383.
13. Heyworth, R.; Walker, P. G. *Biochem. J.* **1962**, *83*, 331-336.
14. Arai, S.; Suzuki, H.; Fujimaki, M.; Sakurai, Y. *Agric. Biol. Chem.* **1966**, *30*, 263.
15. Sessa, D. J.; Warner, K.; Rackis, J. J. *J. Agri. Food Chem.* **1976**, *24*, 1621.
16. Huang, A. S.; Hsieh, O. A. L.; Chang, S. S. *J. Food Sci.* **1982**, *47*, 19-23.
17. Barnes, S.; Coward, L.; Kirk, M.; Sfakianos, J. *Proc. Soc. Exp. Biol. Med.* **1998**, *217*, 254-262.
18. Kao, C. H. Master thesis, National Chung Hsing University, Taiwan., **1998**.

Chapter 7

Changes in Functional Factors of Sesame Seed and Oil during Various Types of Processing

Mitsuo Namiki[1,6], Yasuko Fukuda[2], Yoko Takei[3],
Kazuko Namiki[4], and Yukimichi Koizumi[5]

[1]Nagoya University, Nagoya, Japan 464–8601
[2]Shizuoka University, Shizuoka, Japan 422–8529
[3]Osaka Kyouiku University, Osaka, Japan 582–8582
[4]Sugiyama Jyogakuen University, Nagoya, Japan 464–8662
[5]Tokyo University of Agriculture, Tokyo, Japan 156–8502
[6]Current address: Meitoku Yashirodai 2–175, Nagoya, Japan 465–0092

Sesame seed and oil have long been used as a representative health food, and recently, various important physiological activities of sesame lignans have been elucidated. Most sesame foods are produced by roasting at about 150 ℃ to develop characteristic flavor and taste. The sesame oil from seeds roasted at 180-200 ℃ have a characteristic flavor and brown-red color, and are very stable against oxidative deterioration. Sesame salad oil from unroasted seeds is commonly purified, and is also stable against oxidation. Among lignans, sesamin was stable in roasting and almost no change occurred even at 200℃, while sesamolin decomposed mostly to give sesamol, especially in deep frying. The marked antioxidative activity of deep-roasted oil was shown to be caused by the multi-synergistic effects of Maillard-type roasted products, γ-tocopherol, sesamol, and sesamin. A very interesting fact is that sesamolin was changed effectively to sesaminol, a newly discovered antioxidative lignan, during the decolorization process of unroasted sesame oil. The deep-roasted sesame flavor concentrates containing various alkylpyrazines showed marked antithrombosis activity. Sesame lignans, antioxidative factors, and also characteristic flavor components could be extracted specifically by supercritical CO_2 extraction from sesame seed or oil.

Sesame seed and oil have long been used worldwide as a representative health food. Old Egyptian records extol its merits as a source of energy, and Hippocrates noted its high nutritive value. The magic words "open sesame" demonstrated its popularity in Arabic countries. Old Chinese literature tells us of its daily use to control body action and prevent senility (1,2).

However, until recently there were no scientific studies to elucidate these interesting physiological activities of sesame. In recent years, beginning with our studies on antioxidative factors contributing to highly stable characteristics of sesame oil for oxidative deterioration, many Japanese scientists investigated its physiological activities extensively and elucidated various interesting functionalities due mainly to characteristic sesame lignans, such as sesamin, sesamolin and a new lignan, sesaminol (1-3).

This paper concerns the development and enhancement of these functionalities in the course of various processes and cooking of sesame foods and oils such as roasting, supercritical fluid extraction and purification of oil.

General and Functional Components of Sesame Seed

A cultivated species of sesame, *Sesamum indicum* L., is a major commercial source of sesame and is grown in India, China, Sudan, Mexico, and elsewhere. The seeds vary considerably in color, from white through various shades of brown, gray, gold and black. The seed coat may be either rough or smooth, are tiny and weigh 2-3,5g/1000 seeds (1,4).

The major constituents of sesame are oil, protein and carbohydrate, and their content differ somewhat depending on the variety (2,5). Sesame is a high-energy food containing aproximately 50% oil, consisting mainly of oleic acid and linoleic acid, with small amounts of palmitic and stearic acid but with only trace amounts of linolenic acid (2). Oleic acid and linoleic acid are common fatty acids in food, and linoleic acid is one of the essential fatty acids. According to recent studies in fatty acid nutritional science, intake of n-3 fatty acid (e.g., linolenic acid) is recommended (3).

Sesame oil is known to contain very little trans fatty acids (2). Thus, sesame oil is considered to be good in nutritive value, although it is not that much superior oils such as soy bean oil.

Sesame contains approximately 20% protein. Compared with the standard values recommended by the FAO and WHO, sesame protein is slightly lower in lysine but richer in other amino acids, especially methionine, cystine, arginine and leucine, so its nutritional value is considerable. Good growth in rats resulted when it was tested as a mixture with soybean protein which is rich in lysine but low in methionine *(1,2)*.

Carbohydrates comprise of about 18-20% and contain small amounts of glucose and fructose, but no starch is present. They seem to be present mostly as dietary fiber *(2)*. Sesame contains significant amounts of the vitamin B group, although it can not be considered to be a major source of B vitamins in the diet because of its low consumption. Sometimes sesame is said to be rich in vitamin E as an important factor contributing to a healthy food. However, the tocopherol found in sesame is largely γ-tocopherol, and the α-tocopherol content is very small. The vitamin E activity of γ-tocopherol is evaluated to be about 5% that of α-tocopherol, so sesame must be considered poor in vitamin E *(2)*.

Sesame is rich in various minerals especially in calcium and iron and also notable for its selenium content. Based on these nutritional investigations, sesame is high quality food comparable to soybean, and rice. To better explain the traditionally evaluated physiological effects of sesame seed and oil noted above, further studies have been conducted.

Sesame Lignans and their Functional Activities

Our research group noted especially rich minor components of sesame, sesame lignans such as sesamin, sesamolin and others.   Starting with our chemical researches on the extraordinarily strong antioxidative stability of oil obtained from roasted and unroasted sesame seeds, many studies on the chemistry and physiological activities of sesame lignans have been reported in Japan.

Sesamin and sesamolin are known as representative and characteristic sesame lignans, their contents are usually 0.5-0.8% and 0.3-0.5%, respectively, in seed. Sesaminol and its glucosides, sesamolinol, pinoresinol, and P-1 have been newly isolated and identified by our group as antioxidative sesame lignans *(2,3)*. Changes of sesamolin to sesaminol and sesamol during food processing are very important for the development of marked increase in antioxidative activity. This is explained in a later section.

Various important and interesting functional activties of sesame seed and lignans investigated by Japanese researchers *(2,3)*. Sesame lignans, especially sesaminol, showed marked antioxidative activity not only in vitro, but also in vivo, e.g. long-term feeding of sesame   showed clean suppression of aging marked points of senescence accelerated mice (SAM) *(2,6)*. Concerning this antiaging effect it was demonstrated that sesame and sesaminol suppress significantly increase of lipid peroxidation in liver, red cell hemolysis, and plasma pyruvate kinease *(7)*. These effects demonstrate that sesame lignans have strong synergistic and enhancing effects on vitamin E activity of tocopherols *(8)*. Suppression of LDL lipid peroxidation was also observed with sesaminol glucosides *(9)*.

Specific inhibitory effect on $\triangle 5$ desaturase in fat metabolism by sesame lignans was first found in studies on the microbial production of highly unsaturated fatty acids *(10)*, and extended in animal experiments, which demonstrated many important effects on fatty acid metabolism such as modification of the fatty acid profile of liver phospholipids *(11)*, contribution to maintaining n-6 and n-3 fatty acid balance *(12)*, and acceleration of oxidative metabolism of fatty acid *(13)*. Sesame lignans were shown to enhance liver functions and metabolism of alcohol *(14)*, and also to have hypochoesterolemic acitivity *(15)*, immunoregulatory functional activity *(16)*, prevention of chemically induced cancer *(17)*, and others.

These interesting activities of sesame lignans strongly substantiate the traditionally believed medicinal functions of sesame, and drew much attention to lignans as a unique group of functional factors of health foods, making them comparable to polyphenol groups and carotenoid groups. In this paper, we discuss the development and changes in these functional activities of sesame mainly due to its lignans during various types of food processing, especially on roasting and supercritical carbon dioxide extraction.

Processing of Sesame Food and Oil

In North American and European countries, sesame is used mainly as a topping on bread and biscuit, but in Japan, Korea, China and other Asian countries, there exist a variety of sesame foods and oils as shown in *Fig.1 (1-3)*.

The most important and characteristic processing of these sesame foods and oils, is roasting. In the case of sesame foods, seeds are roasted usually at about 150℃ for 10-15 min, which develops the characteristic and pleasant sesame flavor and gives various kinds of mashed and paste sesame foods. There are two kinds of sesame oil, deep-roasted oil and unroasted raw oil. The former is used widely in Asian countries, especially in Japan, Korea, and China, as a very important cooking and seasoning oil and deep frying oil as used for Tempura. The seeds are roasted at about 180-200 ℃ for 10-20 min, and the expelled oil is filtered without further purification. It has a characteristic brown-red color and roast flavor. The latter, sometimes called sesame salad oil, is prepared by expelling steamed raw seeds followed by common oil purification processes, such as decolorization and deodorization to give a clear and mild flavored oil.

Antioxidative Activity of Roasted Sesame Oil

It has been well known for a long time that sesame oil is highly resistant to oxidative deterioration, e.g., it was used for mummy making in classic Egypt.

Sesame Foods

Oils

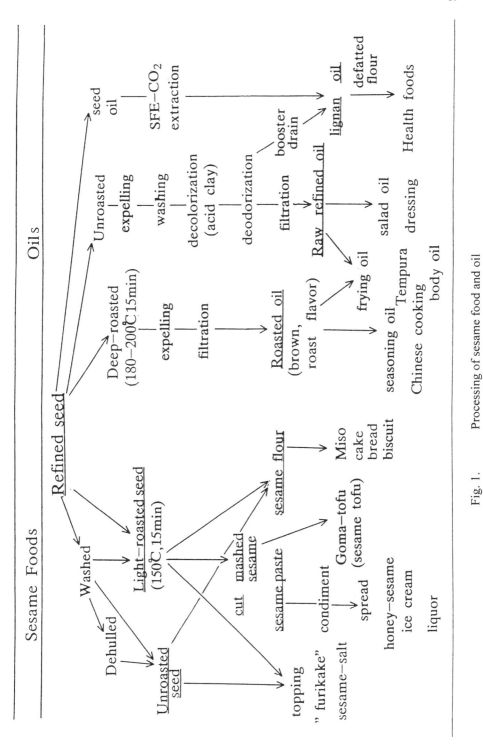

Fig. 1.   Processing of sesame food and oil

Recently, this was demonstrated experimentally, that when common vegetable oils were stored in an open dish at 60 ℃ and autoxidation was determined by increase in weight caused by lipid peroxidation. Soybean oil and others showed rapid increase in weight after about 10 days, while the two types of sesame oil were very stable, and oil of roasted sesame especially showed no increase in weight and hardly oxidized even after storage for 50 days. Unroasted sesame oil was not so stable but significantly more antioxidative than other oils (18).

We then conducted chemical investigation to elucidate the antioxidative factors in these sesame oils. In the case of roasted sesame oil, the antioxidative activity was increased with increase in roasting temperature above 160 ℃ along with an increase in red-browning, indicating the contribution of some roast reaction products in antioxidative activity (19). Isolation of antioxidative factors from deep-roasted oil at 200 ℃ was then conducted by extraction with ether followed by hot methanol. From the ether-soluble fraction, γ-tocopherol and sesamol, decomposition product of sesamolin during roasting and known as antioxidant (20), were identified as the active factors. The methanol-extracted concentrates were separated by XAD-7 chromatography to give a brown antioxidative fraction assumed to be Maillard reactiont products.

The results of antioxidation tests carried out in various combinations of these active fractions demonstrated that the remarkable antioxidative activity of roasted sesame oil might be due to a multi-synergistic effect of the methanol soluble Maillard reaction products + sesamol + γ-tocopherol + sesamin, as shown in *Fig. 2 (21)*.

Antioxidative Activity of Unroasted Sesame Oil

In the case of unroasted sesame salad oil, there are no Maillard-type roasted products and a negligible amount in sesamol, but the oil is highly antioxidative. To investigate antioxidative factors in the unroasted oil, we extracted active fractions with methanol and isolated them by chromatography to discover a new lignan phenol compound named sesaminol as the main antioxidative factor in unroasted sesame salad oil (22). Sesaminol involves sesamol as a moiety and strongly antioxidative comparable to sesamol but far more stable like BHA, probably due to the common structure involving a bulky substituent at ortho position of phenol group. It is found in considerable amounts in the salad oil, but strangely in very small amounts in material sesame seed itself as a free lignan. To elucidate this strange fact, we found marked changes in the lignan content during the purification processes of sesame salad oil, especially in decoloration using acid clay, e.g., there was almost complete loss of sesamolin and formation of appreciable amounts of sesamol, sesaminol and epimers. (*Table I (23)*).

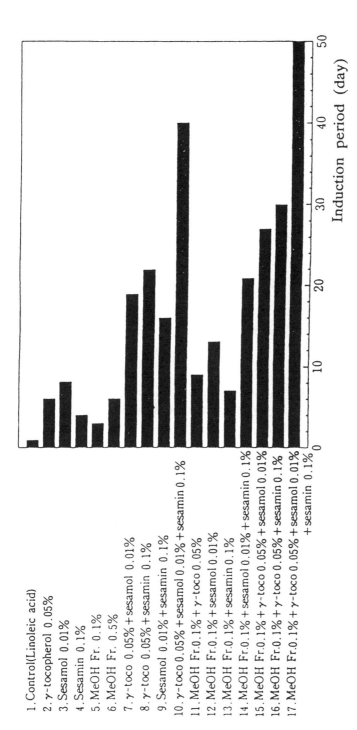

Fig. 2. Synergistic effect of methanol-extracted Maillard reaction products and other factors in antioxidative activity of deep-roasted sesame oil

1. Control(Linoleic acid)
2. γ-tocopherol 0.05%
3. Sesamol 0.01%
4. Sesamin 0.1%
5. MeOH Fr. 0.1%
6. MeOH Fr. 0.5%
7. γ-toco 0.05% + sesamol 0.01%
8. γ-toco 0.05% + sesamin 0.1%
9. Sesamol 0.01% + sesamin 0.1%
10. γ-toco 0.05% + sesamol 0.01% + sesamin 0.1%
11. MeOH Fr.0.1% + γ-toco 0.05%
12. MeOH Fr.0.1% + sesamol 0.01%
13. MeOH Fr.0.1% + sesamin 0.1%
14. MeOH Fr.0.1% + sesamol 0.01% + sesamin 0.1%
15. MeOH Fr.0.1% + γ-toco 0.05% + sesamol 0.01%
16. MeOH Fr.0.1% + γ-toco 0.05% + sesamin 0.1%
17. MeOH Fr.0.1% + γ-toco 0.05% + sesamol 0.01% + sesamin 0.1%

91

Degradative formation of sesamol from sesamolin is known , although most sesamol is lost in the following deodorization process.

The fact that decrease in sesamolin and development of sesaminol occurred at the same step was especially notable, because they have different lignan skeletons. The relationship between these compounds was then studied on the assumption that chemical conversion of sesamolin to sesaminol occurs during decolorization by heating with acid clay. To confirm this assumption, a solution of sesamolin (a) with acid clay in corn oil, and (b) with camphor sulfonic acid in toluene, was heated at 100℃ for 30 min. HPLC analyses of the reaction mixtures demonstrated that sesaminol with its epimers was formed in high yields under anhydrous conditions. Addition of m-cresol to the system (b) produced a new lignan containing the cresol group. A conversion mechanism was proposed involving scission of sesamolin to produce an oxisonium ion and sesamol, and electrophilic addition of sesamol at the ortho position to the oxisonium ion to form sesaminol (*Fig. 3*) *(24)*.

Recently, the presence of considerable amounts of sesaminol as glucosides in a water-alcohol soluble fraction of sesame seed was demonstrated. They are considered as potential factors for the development of functional activities of sesaminol by appropriate processes such as enzyme treatment and acid hydrolysis, and moreover by the activity of intestinal bacteria *(25)*.

Antithrombosis Activity of Deep-roasted Sesame Oil

We have demonstrated that methyl pyrazine groups, such as, 2,3,5-trimethyl pyrazine, have strong antithrombosis activity, as determined by inhibitory effect on human platelet aggregation induced by collagen and measured by turbidimetry as shown in *Table II (26)*.

The deep-roasted sesame oil has a characteristic roast flavor, and it was shown by GC analysis that the volatiles contain many pyrazine compounds *(3)*. Then, we measured the antithrombosis activity of the volatile concentrates of deep-roasted sesame oil collected in water by suction of air around the expeller at a sesame oil factory. As shown in *Table III-A*, the volatile collected water showed the inhibitory activity with its 200x to 400x dilution, and the activity was markedly decreased in the ether extracted residual water. The ether extracts thus obtained at pH 9-10 were about 1.7-2.5g from 1 L of the volatile collected water and based on the GC analysis, the ether extracts at alkaline pH contains much of pyrazine derivatives and estimated to be about 30 % of the extracted matter. The ether extracts, especially at pH 9-10, gave very strong antithrombosis activity as shown in *Table III-B*, indicating there exist much active components involving alkyl pyrazine derivatives in the volatiles collected in water, that is, in the vapor of the deep-roasted sesame oil.

The results suggest that there is some potential to prevent thrombosis by the

Table I. Changes in Content of Sesame Lignans During the Refining Process of Unroasted Sesame Oil (mg/100g oil)

Process	Sesamin	epi-Sesamin	Sesamolin	Sesamol	Sesaminol
Crude oil	813.3	0	510.0	4.3	0
Alkali treatment	730.6	0	458.0	2.5	0
Warm water treatment	677.8	0	424.8	0.7	0
Decolorization	375.5	277.6	0	46.3	81.9
Deodorization	258.3	192.6	0	1.7	62.7

Table II. Antithrombosis Effect of Pyrazine Derivatives

Pyrazine derivative	Inhibitory Activity (mg/ml)				Conc. in
	5.0	0.5	0.1	0.05	sesame flavor
2-Methyl	+++	++	+	-	***
2,5-Dimethyl	+++	++	+	-	***
2,3,5-Trimethyl	+++	+++	+++	+++	***
2,3,5,6-Tetramethyl	+++	+++	+++	++	-
2-Ethyl	+++	+++	+++	++	*
2-Propyl	+++	+++	+++	++	*
2-Acetyl	+++	+++	-	-	***
2,3-Diethyl-5-methyl	+++	+++	+++	+++	***
2-Methyl-3-isopropyl-5-methyl	+++	+++	+++	++	
Aspirin	+++	+++	+++	+++	

Inibitory activity: +++ >50%, ++ 50-30%, + 30-10%, - 10%>
Conc. in sesame flavor (GC); *** high, ** midle, * small

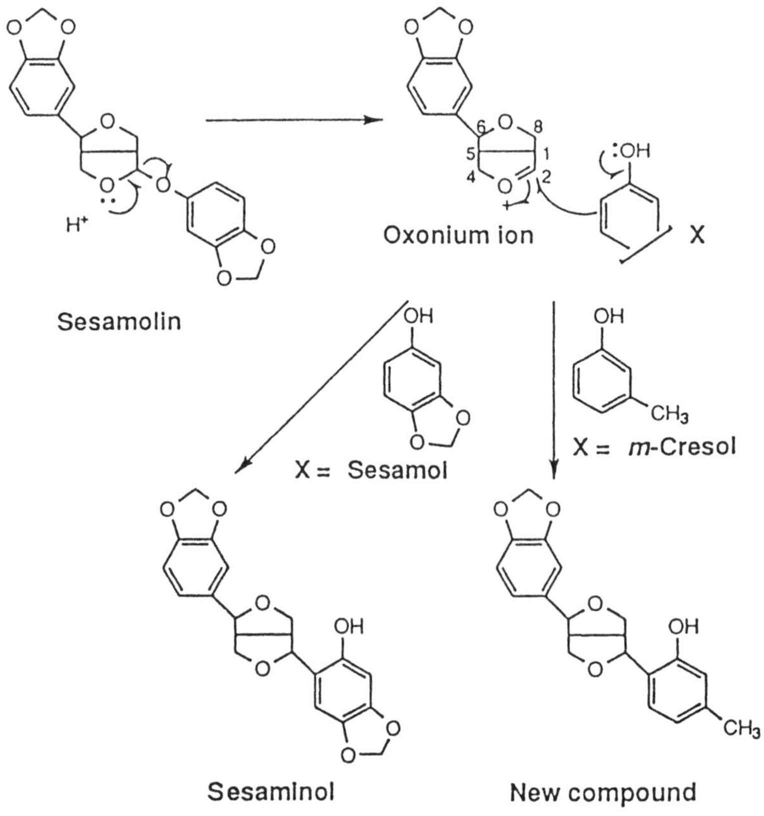

Fig. 3.   Scheme for the mechanism of the formation of sesaminol from sesamolin

Table III. Antithrombosis Activity of the Volatile Components of Deep-roasted Sesame Oil

A (Volatile Collected Water)

Sample	Material Solution	dilution 100X	200X	400X
Volatiles in water	+++	+++	+++	++
Ether extd residues	+++	+++	+	-
Et-acetate extd residues	+++	+++	+	-

Inibitory activity: +++ >50%, ++ 50-30%, + 30-10%, - 10%>

B (Ether Extracts)

Sample	Activity/Final conc. (mg/ml) 0.05	0.01	0.005
Ether extracts (at neutral)	+++	+++	++
Ether extracts (at alkaline)	+++	+++	+++
Ether extracts (at acidic)	+++	++	-
Aspirin (active control)	+++	++	+

Inhibitory activity: +++ > 50%; ++ 50-30%; + 30-10%; -10%>

aromatherapy-like effect of daily intake of pyrazine compounds involved in the flavor of deep-roasted sesame oil, such as found in Japanese tempura.

Supercritical Carbon Dioxide Fluid Extraction (SFE) of Sesame Seed and Oil

Until now, sesame lignans in sesame oil were technically isolated from booster drains containing mostly triglycerides, while minor lignans were isolated during the vacuum deodorization process of unroasted sesame oil. In this process, either sesamin or sesaminol was obtained in a nearly half-and-half mixture of their native and epi-forms, as shown in *Table 1 (23)*.

Recently, supercritical fluid extraction technique (SFE) has come into use as a very clean and safe method of oil extraction without producing residual organic solvent such as n-hexane and with less contamination by phospholipids. Also, SFE is sometimes used to extract some important specific components such as caffeine in coffee and hop resins for beer brewing *(27,28)*. However, there was no SFE experiment on sesame oil, so we tried SFE with carbon dioxide of sesame oil and lignans using the folllowing two methods. A; a small experimental apparatus of 300 ml (NOVA Co., Switzerland), 300-350 Bar and 1.2 kg/h at 40℃. B; a large-scal plant of 500 l (Krupp Co., Germany), 150-350 Bar and 1,000 kg/h at 40 ℃.

The time-course of extraction of oil and lignan from the light-roasted sesame seed with n-hexane and by supercritical carbon dioxide fluid extraction (SFE-CO_2) is shown in *Fig.4*. In the case of extraction with n-hexane, oil and lignan in sesame seed were extracted almost in parallel and could not be separated from each other, while by the SFE-CO_2, lignans were extracted much faster than oil, as shown in B, indicating that earlier fractions of extracted oil contain higher concentrations of lignans.

We next performed extraction of lignans and other oil-soluble minor components from sesame oil by SFE(CO_2).

Figure 5 shows the extracted amounts of sesamin and sesamolin, the main lignans in sesame, and γ-tocopherol, the main tocopherol in sesame, in the fractions with time course of the SFE from roasted sesame oil. Extraction was performed by the large-scale plant and concentrations of lignans and tocopherol were determined by HPLC. As shown in the figure, most lignans in sesame oil were extracted in earlier fractions within the first 2 hours giving very high concentrations as indicated on the logarithmic scale of the abscissa. The extracted volume of the oil in the earlier fractions was usually about 10-15 % of the material oil, though it depends on the pressure and other extracting conditions, the lignans extracted in these fractions were concentrated to about several to ten times of that of the material. Then, the lignans in them were sometimes saturated (ca. 3 %) and partially precipitated in a crystalline form.

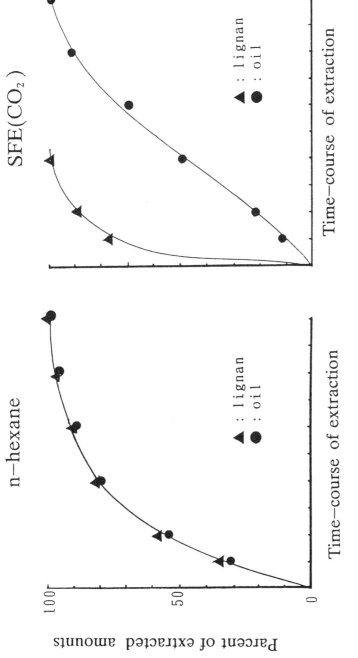

Fig. 4.   Time-course of extraction of oil and lignans from sesame seed by n-hexane and SFE

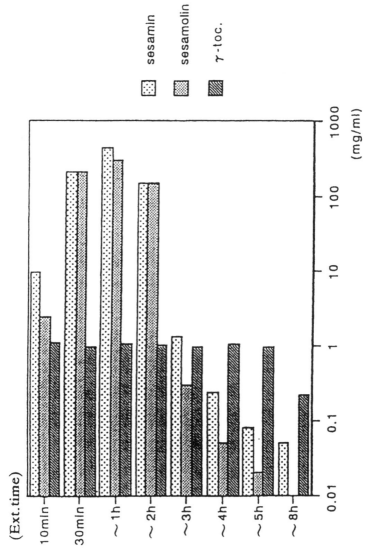

Fig. 5. Extracted amounts of lignans and γ-tocopherol in SFE fraction of roasted sesame oil

Here, sesamin was extracted slightly faster than sesamolin, and moreover sesamin is less soluble than sesamolin in the oil, so the precipitated lignan in the earlier fractions appear to consist mainly of crystalline sesamin.

A notable fact is that the lignans thus extracted consist only in the native form and not in the epi-form. This is different from lignans obtained from the booster drains of the deodorization process of unroasted oil.

It was also noted that γ -tocopherol was extracted almost constantly during the course of extraction but gave no concentrates.

Antioxidative Activity of Fractions of SFE of Roasted Sesame Oil.

Figure 6 shows the oxidative stability of the SFE fractions of roasted sesame oil tested by the weighing method where by increases in the weight of a given amount of oil in an open Petri dish stored at 70℃ were indicated. Increase in weight means increase in lipid peroxidation.

As can be seen, the earlier fractions of SFE were very stable while the later fractions were weak in stability, indicating that the antioxidative factors in the roasted sesame oil were extracted rapidly by SFE and concentrated in the earlier fractions.

As noted above, γ -tocopherol is assumed to be one of the main antioxidative factors in roasted sesame oil, and it was extracted constantly during SFE but did not concentrated, so the concentrated antioxidative factors may not include tocopherol and may be those of lignan groups.

Extraction of the Characteristic Flavor of Roasted Sesame Seed by SFE

In GC analysis of the volatile head space gases in the earlier and the later SFE fractions of light-roasted and deep-roasted sesame seeds, it was demonstrated that the earlier fractions gave far more peaks than the later fractions, and large peaks could be observed with very small retention times, especially in the case of light-roasted sesame oil. Compared with the results determined by GC-MS analysis in our previous studies *(3, 29-32)*, the peaks observed at a very early time were assigned to be low molecular aldehydes, alcohols and sulfur-containing compounds and the following peaks were those of various pyrazines and pyrrols, suggesting that the characteristic roast sesame flavor may be concentrated in the earlier fractions of SFE.

This characteristic of SFE was supported by sensory test analysis for each fractions of SFE of roasted sesame oil. The test was carried out by fifteen panelists of the college women on odor intensity score of 1 to 5, from weak to strong, and rank of aroma preference of 1 to 4, from most preference to least preference. *Table IV* shows the results obtained on the light-roasted sesame seed, the earlier fractions of SFE were highly evaluated in sesame-like flavor either in

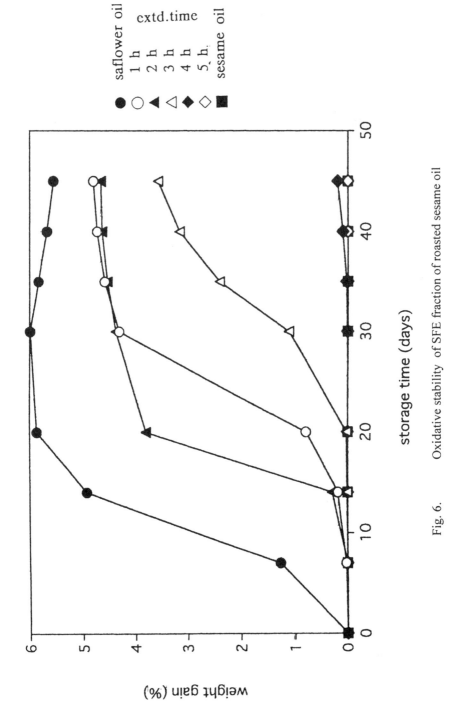

Fig. 6.   Oxidative stability of SFE fraction of roasted sesame oil

Table IV. Sensory Evaluation of Flavor of SFE Oil Fractions from Light-
roasted Sesame Seed

| Evaluated Attribute | Odor Intensity Score and Rank Sum of Aroma Preference(#) | | | |
	Fr.1(~0.5h)	Fr.2(~1.0h)	Fr.3(~2.0h)	Fr.4(~4.0h)
Sesame-like	4.5±1.0a (22*)	3.5±1.1a (28)	2.2±1.1b (40)	1.9±1.2b(50*)
Peanut-like	3.1±1.4a,b (30)	3.6±1.2a (25*)	3.1±1,a,b (38)	2.1±1.3b(47*)
Oily-odor	3.5±1.4a(27)	2.9±1.3a (30)	3.2±1.1a (33)	3.1±1.4a (40)
Off-flavor	2.4±1.3a(23*)	2.3±1.4a (26)	2.6±1.4a (34)	2.8±1.5a (41)
Overall	4.5±0.5a(27)	3.5±1.1b (28)	2.9±1.0b (34)	3.3±1.0b(49*)
Preference	3.5±1.1a	3.5±1.2a	2.8±1.0a,b	2.5±1.1b

a,b; means with different letters in the same line are significantly different
at the 5% probability
* ; significant difference at the 5% probability level in the same line
Intensity score; 1 - 5 (weak to strong). mean values of 15 panelists.
Rank of preference; 1 (most preferred) to 4 (least preferred), in (#) sum of
rank number of 15 panelists.
SFE: Supercritical carbon dioxide fluid extraction from sesame seed at 350 Bar

the mean intensity score and the rank sum of preference. The results indicate that SFE-CO$_2$ is a very effective and important method for the extraction and concentration of the characteristic sesame flavor.

Discussion and Conclusion

Sesame has long been regarded as representative health food which increases energy and prevents aging, and sesame oil has been known empirically as being highly resistant to oxidative deterioration. Recently, these effects were elucidated scientifically to be due mainly to the antioxidative and physiological effects of sesame lignans.

Among various processes used to produce a variety of sesame foods and oils, the most important is roasting, which produces the characteristic flavor, taste and color of sesame foods. Moreover, roasting has been shown to induce strong antioxidative activity in oil. This in vitro activity was elucidated to be due to the multi-synergistic effects of Maillard-type roasting products + γ-tocopherol + sesamol (degradation product of sesamolin by roasting) + sesamin, although there remained important problems to be elucidated. Further investigation is needed to know what kind of chemical components in the Maillard-type reaction products are effective with what type of reaction mechanism in the antioxidative or synergistic effect, e.g., presence of enaminol structure acting as reductive radical scavenger or chelating agent.

Concerning functional and physiological problems, it is necessary to investigate whether the marked antioxidative activity of the deep-roasted sesame oil in vitro is also effective against in vivo oxidative damage when it is ingested as a food constituent.

In this respect, it is notable that marked antithrombosis activity was observed on the roast sesame flavor probably due to the involvement of various pyrazine derivatives which have been demonstrated to be effective in antithrombosis activity.

The roasting of food and other materials is a widely used process to develop characteristic flavor, taste and color, and is sometimes important as a key step in determining food quality, as in the case of coffee. The results shown here demonstrate that roasting plays important roles not only in the formation of the sensory qualities of food but also in the development of functionalities such as antioxidative and antithrombosis activities. Studies examining this new aspects should be conducted on various roasted foods.

The significantly high antioxidative activity of unroasted sesame oil was demonstrated to be due to a newly identified antioxidative lignan phenol, sesaminol. Very interestingly, this sesaminol is produced by the intermolecular rearrangement reaction from sesamolin catalyzed by acid clay during the decolorization process of unroasted raw sesame oil. This new finding that a marked functional activity of food is developed during a common food process by nobel reaction is notable.

Supercritical fluid extraction (SFE) is a clean and safe oil extraction technology and a sometimes very powerful tool in the extraction of special components. It was demonstrated that SFE-CO_2 of sesame seed and oil induced specific extraction and condensation of sesame lignans, antioxidative factors and characteristic flavors as their native forms. The fact that the SFE provides crystalline native lignans is especially important in studies on the functional activity of sesame lignans, because until now most studies on the physiological activities of sesame lignans as listed were conducted using a nearly half-and-half mixture of sesamin with its epimer, and though not yet determined, the functional activities of the epimer may be inferior to those of native one. So, by using native lignans isolated by the SFE method it becomes possible to determine exactly the various functional activities of sesame lignans. SFE not only yielding pure lignans, but also provides easily sesame oil concentrate having high functional activities very useful as an important factor in various health foods. Moreover, SFE gave good quality defatted sesame meal which could be further improved by fermentation and enzyme treatments to give highly antioxidative and functional food stuff for the preparation of health foods. For practical utilization of SFE, more cost-effective extraction conditions must be developed.

Acknowledgment

We are grateful to Fujimi Yohouen Co. and Mori Seiyu Co. for collaboration on the SFE experiments, and to Kadoya Seiyu Co. and Kuki Sangyo Co. for supplying sesame materials.

References

1. *Goma no Kagaku (in Japanese) (Sesame Science)*, Namiki,M,;Kobayashi,T., Eds.; Asakura Shoten Co.,Tokyo, 1989, p249
2. Namiki,M., *Food Reviews International*, **1995**, *11*, 281-329
3. *Goma, sono Kagaku to Kinousei(in Japanese)* (Advances in sesame science and function), Namiki,M., Ed.; Maruzen Planet Co., Tokyo, 1998, p268
4. *Oilseeds Crops* Joshi,A.B., Longman, London, 1983, p282
5. *Standard Tables of Food composition in Japan*, Ed., Scientific Technique Agency of Japan, 1982
6. Yamashita,K.;Kawagoe,Y.;Nohara,Y.;Namiki,M.;Osawa,T.;Kawakishi,S., *Eiyo Shokuryo Gakkaishi*, **1990**, *43*, 445-449
7. Yamashita,K.;Nohara,Y.;Katayama,K.;Namiki,M., *J. Nutrition*, **1992**, *122*, 2440
8. Yamashita,K.;Iizuka,Y.;Imai,T.;Namiki,M., Lipids, 1995, 30, 1019-1028

104

9. Kang,M-H.;Naito,M.;Sakai,K.;Uchida,K.;Osawa,T.   *J. Nutr.*, **1998**
10. Shimizu,S.;Akimoto,K.;Kawashima,H.;Shinmen,Y.;Yamada,H., *JAOCS*, **1989**, *66*, 237-241
11. Shimizu,S.;Akimoto,K.;Shinmen,Y.;Sugano,M.;Yamada,H., *Lipids*, **1991**, *26*, 512-516
12. Umeda-Sawada,R.;Ogawa,M.;Okada,Y.;Igarashi,O., *Biosci.Biotech. Biochem.*, **1996**, *60*, 2071-2072
13. Umeda-Sawada,R.;Ogawa,M.;Igarashi,O., *Lipid*, **1998**, *33*, 567-572
14. Akimoto,K.;Kitagawa,Y.;Akamatsu,T.;Hirose,N.;Sugano,M.;Shimizu,S.; Yamada,H.,   *Ann. Nutr. Metab.*, **1993**, *37*, 218-224
15. Hirose,N.;Inoue,T.;Nishihara,K.;Sugano,M.;Akimoto,K.;Shimizu,S.; Yamada,M.,   *J. Lipid Res.*, **1991**, *32*, 629-638
16. Nonaka,M.;Yamashita,K.;Iizuka,Y.;Namiki,M.;Sugano,M., *Biosci. Biotech. Biochem.*, **1997**, *64*, 836-839
17. Hirose,N.;Doi,F.;Akazwa,K;Chijiiwa,K.;Sugano,M.;Akimoto,K.; Shimizu,S.;Yamada,H., *Anticancer Res.*, **1992**, *12*, 1259-1266
18. Fukuda,Y.;Osawa,T.;Kawakishi,S,;Namiki,M., *Nippon Shokuhin Kagaku Kogaku Kaishi*, **1988**, *35*, 28-32
19. Koizumi,Y.;Fukuda,Y.;Namiki,M., *Nippon Shokuhin Kagaku Kogaku Kaishi*, **1996**, *43*, 689-694
20. Budouski,P.;Menezes,F.G.T.;Dollear,F.G., *JAOCS*, **1950**, *27*, 377
21. Fukuda,Y.;Koizumi,Y.;Ito,R.;Namiki,M., *Nippon Shokuhin Kagaku Kogaku Kaishi*,   **1996**, *43*, 1272-1277
22. Fukuda,Y.;Osawa,T.;Namiki,M.,   *Agric.Biol.Chem.*, **1985**, *49*, 301-306
23. Fukuda,Y.;Nagata,M.;Osawa,T.;Namiki,M.,   *JAOCS*, **1986**, *63*, 1027-1031
24. Fukuda,Y.;Isobe,M.;Nagata,M.;Osawa,T.;Namiki,M.,   *Heterocycles*, **1986**, *24*, 923-926
25. Kang,MH.;Kawai,Y.;Naito,M.;Osawa,T.,   *J.Nutr.* **1999**,*129*, 1885-1890
26. Igarashi,M.;Kanno,H.;Tanaka,K.;Asada,T.;Yamaguchi,R.;Shirai,T.; Yamanaka,M.; Namiki,K.; Namiki,M., *J.Med.Soc.Toho,Japan*, **1986**, *33*, 261-264
27. *Supercritical Fluid Processing of Food and Biomaterials*, Rizvi,S.S.H., Blackie Academic & Professional, Glasgow, **1994**
28. Eller,F.J.;King,J.F., *J. Agric. Food Chem.*, **1998**, *46*, 3657-3661
29. Takei,Y., *J. Home Economics of Japan*, **1988**, *39*, 803-815
30. *Goma no Kagaku*; . Takei,Y.;Nishimura,O.;Mihara,S.,Eds. Namiki,M.. Kobayashi,T., **1989**, 143-155
31. *Goma no kagaku to Kinousei*; . Takei,Y., Ed. Namiki,M. **1998** 124-131
32. Asai,Y.;Fukuda,Y.;Takei,Y.. *J. Home Economics of Japan*, **1994**, *45*, 279-287

Chapter 8

Food Processing Reduces Size of Soluble Cereal β-Glucan Polymers without Loss of Cholesterol-Reducing Properties

Wallace H. Yokoyama[1], Benny E. Knuckles[1], Delilah Wood[1], and George E. Inglett[2]

[1]Western Regional Research Center, Agricultural Research Service, U.S. Department of Agriculture, Albany, CA 94710
[2]National Center for Agricultural Research, Agricultural Research Service, U.S. Department of Agriculture, Peoria, IL 61604

Grinding and shear degrade cereal β-glucans but cooking and starch hydrolysis increase the extractability and availability of the polysaccharide to interact with digesta in the intestinal lumen. Cereal β-glucans in oat and barley are easily degraded by grain milling machinery and endogenous β-glucanases as shown by size exclusion chromatographic separation of soluble polysaccharides and molecular weight and size characterization by multiple angle laser light scattering. Cooking in water or alkaline solutions extracts β-glucans found mainly in the cell walls of oat and barley grains. Shear generated by stirring or pumping is shown to degrade β-glucans. Animal studies demonstrate that β-glucans that have been reduced in mass by an order of magnitude are still able to reduce plasma cholesterol. Moreover, molecular mass of β-glucans from the stomach and intestine of animals are even lower than the size reduced fiber from processing. These results suggest that overall food processing improves the activity of β-glucans by increasing availability and that polymer molecular mass, while reduced, is still effective in plasma cholesterol reduction.

Introduction

Mixed linkage, β-glucan, a form of soluble fiber, found in oat, barley and in small quantities in other cereal grains reduces plasma cholesterol levels in humans and may have other healthful properties (1). High plasma cholesterol is a risk factor for cardiovascular disease. Whole oat products containing β-glucans have been awarded the right to claim health benefits related to cardiovascular disease by the U.S. FDA. β-glucans also reduce the postprandial rise of glucose and insulin, suggesting a role in diabetes management (2). However, the physiological properties of cereal β-glucans are dependent on their bioavailability and molecular characteristics. Our studies using multiple angle laser light scattering to determine polymer size show that endogenous enzymes and mechanical processes reduce polymer size but that cooking increases bioavailability. Animal studies indicate that the limited polymer size degradation that occurs in food processing does not affect plasma cholesterol lowering properties.

β-Glucan Extraction and Characterization by SEC/MALLS

Sample preparation for molecular mass and size analysis by size exclusion chromatography (SEC) and multiangle laser light scattering (MALLS) detection has been described previously (3). Generally, barley or oat kernels were stabilized against enzymatic degradation of β-glucan by heat and/or aqueous alcohol treatment. The kernels were ground into flour and passed through a 0.1 mm screen. Usually, 50 mg of flour was extracted into 1.0 mL of either cold or hot water, or 0.1 N NaOH. The extracted solution, containing a mixture of soluble plant polysaccharides, proteins and other polymers, was filtered through a 0.4 μm filter and 50 μL injected into a high performance liquid chromatograph. The chromatographic system consisted of four size exclusion columns suitable for the molar mass range $1 \times 10^4 - 5 \times 10^7$ g/mol (Waters, Milford, MA), refractive index detector, MALLS (Dawn DSP-F, Wyatt Technologies, Santa Barbara, CA), fluorescence detector, in-line postcolumn calcofluor reactor, and a diode array detector. A schematic of this system is shown in Figure 1. The formation of the fluorescent calcofluor complex with β-glucan allows for unambiguous assignment of the eluting β-glucan peak.

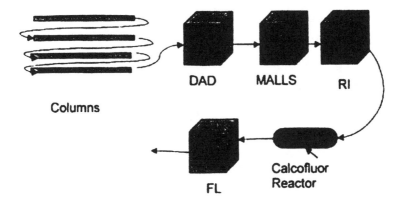

Figure 1. Schematic of SEC/MALLS system used for the separation and molecular weight determination of cereal β-glucan. Four size-exclusion columns in series separate soluble polysaccharide polymers extracted from cereal flours. Mw is determined by multiple angle laser light scattering (MALLS) in combination with RI. Positive identification of β-glucan is possible by in-line reaction of β-glucan with calcofluor and detection of the fluorescent complex.

β-Glucan Characteristics and Location in the Plant Cell

β-glucans are linear, (1→3) and (1→4) mixed linkage, β-d-glucan polymers, and are structural components of the cell walls of oat and barley. The fluorescence micrograph of a barley kernel cross section (Figure 2) is illustrative of the β-glucan distributed in the cell wall of cereal grains. Fluorescence is observed by the formation of the complex of β-glucan with the whitener calcofluor. In barley, β-glucan is uniformly distributed throughout the kernel as shown by fluorescence micrograph (Figure 2) whereas oats have a higher concentration of β-glucan in the outer or bran layers. Therefore, pearling or the removal of the outer layers, a common milling practice for barley does not decrease the β-glucan concentration. However in oat, β-glucan concentration is higher in the outer (or bran) layers of the kernel so that the bran fraction is usually higher in β-glucan.

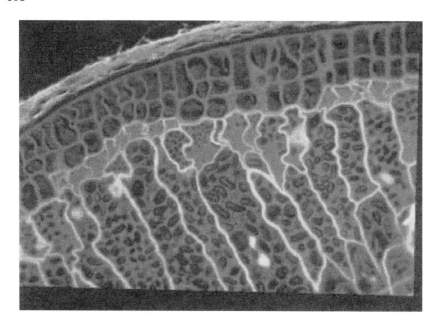

Figure 2. Micrograph of barley kernel section stained with calcofluor results in fluorescent complex of β-glucan. β-glucan only occurs in the cell wall. In barley β-glucan is uniformly distributed in the endosperm.

The weight-averaged molecular weight (Mw) of β-glucans in oat and barley are typically in the 1-3 million range (*3-5*). The solutions formed by as little as 1% β-glucan are extremely viscous and typical of long, linear, soluble polymers. Although β-glucans are mainly soluble polysaccharides, not all the β-glucans in oat or barley can be extracted completely by cold or hot water from barley flour (Figure 3). Aqueous NaOH is required to completely extract β-glucans. However in aqueous NaOH, β-glucans decompose unless stabilized by 1% NaBH$_4$. Sequential extraction starting with cold water, followed by hot water and finally by 0.1 N NaOH results in fractions with decreasing Mw (Figure 4). The Mw of β-glucan is higher from cold water extracts of flour extracted than from either hot water extracts or aqueous NaOH. Similar results for sequential extraction and Mw distribution of the extracted fractions have been reported by Bhatty (*6*) and Beer et al (*5*) for a variety of oat and barley cultivars. The reduction of Mw in later extracts by β-glucanase is always a possibility, but the results of this and other studies suggest that most soluble β-glucans are high in Mw. Extrusion cooking increased the amount of soluble β-glucans. The solubility of β-glucans can also be affected by the distribution of (1→3) and (1→4) glycosidic linkages (7). Insoluble β-glucans with only β-(1→3) linkages are found in yeasts and other organisms, but only minor quantities occur in barley (8).

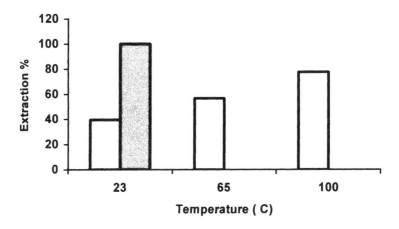

Figure 3. Extraction (%) of β-glucan by water and 1 N NaOH. The extraction of β-glucan by water (unshaded bars) alone increases with temperature but is not complete even at 100 °C compared to extraction with 1 N NaOH (shaded bar). Cooking in water and mechanical food processes may increase extraction and bioavailability.

Although barley and oats are often used in foods in whole or pearled forms, they can be substituted for part of the wheat flour commonly used in baked and/or extruded products. Grinding of barley kernels into flour results in degradation of the β-glucan polymer as shown in Table I, due to the shearing action of the mill. The Wiley mill has a rapidly, rotating knife edge that reduces grain size by a cutting action which causes less relative shear than a Udy Mill which grinds the grain against an abrasive ring. The Mw of the β-glucan prepared from the Udy mill is about 50% of that for the polymer prepared from the Wiley mill. It is always important to bear in mind that in addition to β–glucan polymer size reduction by mechanical shear, β-glucanase activity may also contribute to differences in Mw; perhaps more shearing action releases more β-glucanase enzymes or creates more surface area for the enzyme on which to react.

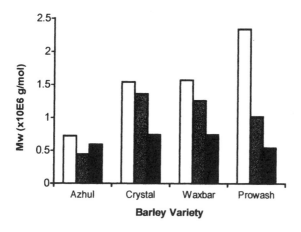

Figure 4. Mw of β-glucan fractions from sequential extraction of flours from four varieties of barley. The first extraction by cold water (unshaded bar) extracts β-glucans with the highest Mw. Mw of β-glucan decreases with subsequent extractions by hot water (medium shade bar) and 1 N NaOH with 1% NaBH₄ (dark shade bar).

Table I. Effect of Grinding on Barley β-Glucan Mw

Variety	$Mw \times 10^5$ (g/mol)	
	Wiley Mill	*Udy Mill*
Steptoe A	2.53	1.11
Steptoe B	3.00	2.28
Azhul	3.64	1.79

If β-glucans are employed in fluid food processes, such as in the production of oat flour in Oatrim and the subsequent use of Oatrim as a fat replacer in ice creams, puddings, salad dressings, or other fluid foods, shear will degrade β-glucan polymers during mixing or pumping. Table II shows the effects of slow and fast magnetic mixing, vortex mixing and syringe injection of a 1 mg/mL β-glucan solution. The Mw of β-glucan forced rapidly through a 20 gauge syringe needle is about 50% that of β-glucan from slow magnetic stirring. Viscosity and correlated Mw decrease has also been reported during the pilot plant scale production of oat β-glucan (9).

Table II. Effect of Shear on β-Glucan Mw in Solution

Sample	Mw x 10^5 (g/mol)			
	Magnetic Mixer, Slow	Magnetic Mixer, Fast	Vortex Mixter	Syring Needle
β-Glucan 1	5.47	3.80	--	--
β-Glucan 2	7.79	5.81	--	--
65°C Extract	10.40	--	6.21	4.91

Another potentially important area of oat and barley flour use is in yeast leavened-bread. Bread was prepared by substituting β-glucan enriched barley flour for 20% of the wheat flour. Bread was prepared from this flour mix using the AACC basic straight-dough fermentation method (10). ß-glucans were extracted with 1 N NaOH and analyzed by SEC/MALLS. At 0 min a single peak eluting at 24 min. was observed. Between 25-26 min the peak became biphasic and broadened to elute between 24-34 min. The Mw of the 25 °C and 100 °C soluble fractions are shown in Figure 5. Enzymes present in the barley flour, added malt or yeast have presumably caused β-glucan polymer degradation since most of the size decrease occurs after mixing.

Animal Studies of Plasma Cholesterol Lowering by Processed Oat

During food processing endogenous β-glucanases and mechanical forces degrade β-glucan polymers. However, unlike chemical reactions of small molecules where entirely new compounds are formed, the degradation of β-glucan polymers by hydrolysis yields polymer products that are smaller in size but essentially chemically similar to the parent compound. For example, the interruption of a 2 million g/mol polymer chain at 3 places can produce four polymer fragments of 500,000 g/mol. Viscosity is significantly decreased by the decrease in polymer length but the products are still β-glucan polymers. The effectiveness of low Mw β-glucan polymers in the reduction of plasma cholesterol was investigated in animal studies.

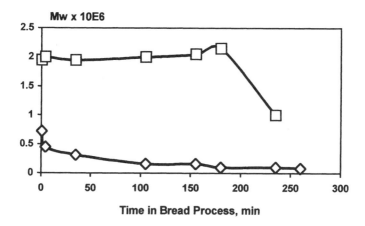

Figure 5. Mw of β-glucan changes with time during a standard breadmaking process. The cold water soluble β-glucans (◊) are immediately hydrolyzed to lower Mw products but the hot water soluble β -glucans (□) are not reduced until the molding step. Time, Process: 0 min, Start mix; 100 min, First punch; 150 min, Second punch; 175 min, Molding; 235 min, End proof; and 260 min, End bake.

β-glucans and many other natural and synthetic soluble dietary polymers such as guar gum, methylcellulose, and psyllium, reduce plasma cholesterol in humans and animals. Several mechanisms have been suggested including delayed gastric emptying, increased bile acid excretion in the feces, and fermentation into short chain fatty acids in the colon (*1*). Short chain fatty acids have been shown to reduce synthesis of cholesterol *in vitro*. Since all soluble polymers are viscous and the monomeric form of the carbohydrate components show no cholesterol activity there has been an association of viscosity with cholesterol reduction. If greater activity is related to polymer size then perhaps polymer degradation might affect cholesterol lowering properties.

Hamsters fed diets high in fat, particularly saturated fat, and supplemented with cholesterol become hypercholesterolemic within a few weeks. Hamsters were fed hypercholesterolemic diets containing either cellulose, oat flour enriched in β–glucan, or oatrim. Oatrim is a fat replacing ingredient made from amylase treated oat flour. Two experiments were conducted to determine the effectiveness of the additives in lowering cholesterol in hamsters.

In the first experiment, Experiment 1, the composition of the oat ingredients are shown in Table III. The dietary fiber in the oatrim ingredients are almost entirely soluble β-glucan, whereas the oat flour enriched in β-glucan contains about an equal amount of soluble and insoluble fibers. The diets were made up to contain 14-15% total dietary fiber, 10% hydrogenated coconut oil, 5% corn oil and 0.2% cholesterol. Plasma low density lipoprotein (LDL) cholesterol from hamsters fed the oatrim-10 and -5 diets were lowered 42% and 67%, respectively, compared to the control (Figure 6). Total plasma cholesterol was lowered in all the animals fed the oat diets, but was lowest in the diets fed oatrim.

Table III. Oat Ingredient Composition

Component	Oat Bran	Oatrim-10	Oatrim-5
% Solid	91.95	95.01	93.4
Protein, Nx6.25	27.06	10.88	3.44
Fat	2.23	2.50	0.83
Ash	4.69	4.89	2.91
TDF	49.61	11.90	6.90
β-glucans	29.12	10.64	6.36

These results were surprising since analysis of the oat polymers by SEC/ MALLS showed that oatrim products contained β-glucan polymers that were about 40% lower in Mw (Experiment 1, Table IV). The reduction in β-glucan Mw could occur by shear and/or enzymatic hydrolysis as a result of impurities or cross reaction with the starch hydrolyzing enzyme used in the manufacture of oatrim. A laboratory scale simulation of the cooking and hydrolysis of oat flour into a hydrolyzed starch oat product like oatrim was conducted. Cooking alone did not reduce Mw. However, enzymatic hydrolysis reduced Mw about 45%. Shear may also contribute at this stage since hydrolysis of the thick starch paste renders the β-glucan polymer more susceptible to shear.

In experiment 2, the hamsters were fed a diet containing the same amount of total fat as in Experiment 1. Dietary cholesterol was the same (0.2%), however the saturated fat source was different (9% butterfat, 2.5% olive oil, 3.5 % corn oil). Total dietary fiber was 11.7% vs 14-15% in Experiment 1. In this second study, only the LDL/HDL ratios of oatrim and hydrolyzed cooked starch (OBXE) were different (Figure 7) from the control.

114

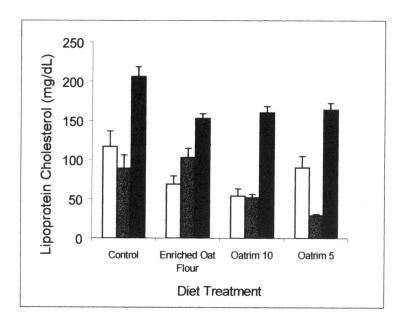

Figure 6. Cholesterol concentration of plasma lipoprotein fractions from hamsters fed raw oat or processed oat diets. Mean LDL cholesterol content (medium shade bars, error bars are SEM) are lower in animals fed oatrim diets compared to either the control or raw oat fed animals. HDL cholesterol (dark bars) is lower in animals fed the three oat diets compared to the control. VLDL cholesterol (unshaded bar) is lower in raw and oatrim 10 fed animals.

Both Experiments 1 and 2 indicate that decreasing β-glucan to 5×10^5 Mw does not decrease the polymers' cholesterol lowering properties. In fact, enzymatic treatment increases its effectiveness at the same dose, suggesting that perhaps other factors such as increased availability at the early stages of the digestive process are important

The β-glucan Mw of the content of the small intestine were determined for the oat diets in Experiment 1 and found to be about 35-50,000 g/mol. Wood et al (4) found that β-glucan from the small intestine of rats had been degraded to about 90,000 g/mol. β-glucans from the small intestine of pigs tended to about 100,000 g/mol (11). Food processing decreases β-glucan Mw, but as this and other studies have shown, they are decreased to a much greater degree regardless of starting Mw by passage through the digestive system.

115

Table IV. Dietary β-Glucan Molecular Characteristics

	Molecular Wt. Mw, g/mol	Molecular Radius, nm	Polydispersity Mw/Mn
Experiment 1			
Oat Bran	1.14×10^6	58.9 ± 1.2	1.35 ± 0.08
Oatrim 10	0.39×10^6	17.2 ± 2.0	2.26 ± 0.04
Oatrim 5	0.477×10^6	17.0 ± 1.5	1.02 ± 0.03
Experiment 2			
Oat Bran (OB)	1.65×10^6	67.1 ± 1.6	1.71 ± 0.15
OB + 100°C (OBX)	1.42×10^6	68.0 ± 1.3	1.41 ± 0.12
OBX + Amylase (OBXE)	9.11×10^5	56.3 ± 1.2	1.36 ± 0.07
Oatrim 10	8.96×10^5	9 ± 2.3	1.73 ± 0.59

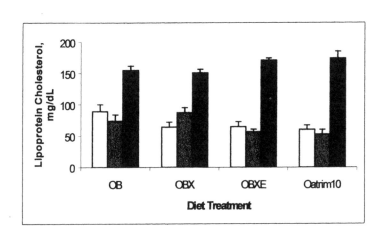

Figure 7. Cholesterol concentration of plasma lipoprotein fractions in hamsters fed raw (OB), cooked (OBX) and enzymatically hydrolyzed (OBXE) oat or oatrim diets. Mean VLDL (unshaded bars), LDL (medium shade bar) and HDL (dark shade bar) cholesterol of the oat fed animals are not different from the cellulose control. The ratio of LDL/HDL is higher in the raw oat fed animals compared to the oatrim fed animals.

References

1. Bell, S.; Goldman, V.M.; Bistrian, B.R.; Arnold, A.H.; Ostroff, G.; Forse, R.A. 1999. *Crit. Rev. Food Sci. Nutri.* **1999**, *39*, 189-202.
2. Wursch, P.; Pi-Sunyer, F.X. *Diabetes Care* **1997**, *20*, 1774-80.
3. Knuckles, B.E.; Yokoyama, W.H.; Chiu, M.M. *Cereal Chem.* **1997**, *74*, 599-604.
4. Wood, P.J.; Weisz, J.; Mahn, W. *Cereal Chem.* **1991**, *68*, 530-536.
5. Beer, M.U.; Wood, P.J.; Weisz, J. *Cereal Chem.* **1997**, *74*, 476-480.
6. Bhatty, R.S. *Cereal Chem.* **1993**, *70*, 73-77.
7. Izawa, M.; Kano, Y.; Koshino, S. 1993. *J. Amer. Soc. Brew. Chem.* **1993**, *51*, 123-127.
8. Fulcher, R.G.; Setterfield, G.; McCully, M.E.; Wood, P.J. *Aust. J. Plant Physiol.* **1997**, *4*, 917-928.
9. Wood, P.J.; Weisz, J.; Fedec, P.; Burrows, V.D. *Cereal Chem.* **1989**, *66*, 97-103.
10. Knuckles, B.E.; Hudson, C.A.; Chiu, M.M.; Sayre, R.N. *Cereal Foods World* **1997**, *42*, 94-99.
11. Johansen, H.N.; Wood, P.J.; Erik, K.; Knudsen, B. *J Agric. Food Chem.* **1993**, *41*, 2347-2352

Chapter 9

Enhanced Iron Absorption from Cereal and Legume Grains by Phytic Acid Degradation

R. F. Hurrell

Laboratory for Human Nutrition, Institute of Food Science, Swiss Federal Institute of Technology Zurich, CH–8803 Rüschlikon, Switzerland

Phytic acid in cereal and legume foods is a potent inhibitor of mineral absorption but can be degraded by activating endogenous phytases during processes such as soaking, blanching, germination and fermentation. Wheat and rye are rich sources of phytases and can be used to completely degrade phytic acid in cereal legume mixtures. Alternatively, commercially available phytases can be used. Iron absorption by human subjects fed a liquid formula meal containing soy isolate was increased 4-5 fold when phytic acid was reduced to zero. Even relatively small quantities of phytate however were strongly inhibitory. Iron absorption was similarly increased 4-10 fold when wheat, maize, rice and oats were dephytinized using a commercial phytase and fed to human subjects as porridge with water. Iron absorption studies in infants using stable isotopes confirmed the beneficial effects of phytate removal from soy formula but not from infant cereals fed with milk.

Phytic acid is a potent inhibitor of mineral absorption. In the gastrointestinal tract, it forms complexes with minerals and with the peptides released on digestion of food proteins. Minerals, such as iron, zinc and calcium, are tightly bound in these complexes and are not completely released for absorption. The absorption of iron, zinc and calcium is thus lower from phytate-containing foods such as cereals and legumes (*1*).

Low iron absorption from cereal and legume-based diets is an important factor in the etiology of iron deficiency in developing countries (*2*) and is particularly important in relation to infant foods, including those manufactured in industrial countries, since infants usually consume a limited number of food items. Young infants, with a rapidly expanding blood volume, have a high iron requirement and without a good supply of bioavailable iron from weaning foods rapidly become iron deficient. Iron deficiency anemia may significantly reduce a child's psychomotor and mental development (*3*), reduce immune status and lead to an increased susceptibility to infection (*4,5*).

Cereal grains and legume products are often used to produce infant foods including weaning products and soy based infant formula. The level of phytic acid in both cereal grains and legume seeds is usually around 1%; in wheat, rice, maize, barley or oats, or in legume seeds such as soybean, lima bean, navy bean and peas (*6*). Milling and bran removal reduce the phytic acid considerably in cereal flours, by 50% in high extraction wheat, 80-90% in low extraction wheat or polished rice and even more in degermed maize. On the other hand with legume seeds, protein concentration or isolation usually maintains or increases the phytic acid level since the phytate is in the protein bodies of the endosperm. A soybean isolate or concentrate may have twice the level of phytic acid as soybean itself.

Phytic Acid and Iron Absorption

Much of the evidence that phytic acid inhibits iron absorption in human subjects has come from the demonstration that degradation of phytic acid during food preparation enhances iron absorption. The degradation of phytic acid in wheat bran by endogenous phytases almost completely removed the inhibitory effect of wheat bran on iron absorption (*7*). Similarly phytic acid degradation by native phytases during bread fermentation and malting of oats increased iron absorption in adults (*8,9*) from these cereals, and using microbial phytases to degrade phytic acid in soy protein isolate increased iron absorption in adults fed soy formula meals (*10*) and infants fed phytate-free soy infant formula (*11*).

Adding phytic acid (7-890 mg) to phytic acid-free wheat rolls inhibited iron absorption in adults in a dose-dependent way (*12*) as did adding increasing amounts of maize bran containing increasing amounts of phytic acid (35-205 mg) (*13*). When soy protein isolate was included as the protein component of a liquid formula meal in adults, the phytic acid level of the isolate had to be reduced from ca. 1% to 0.03% (10 mg phytic acid/meal) so as to achieve a meaningful three fold increase of iron absorption (*10*). The molar ratio of phytic acid to iron had to be reduced from 2.2:1 to 0.1:1 in order to achieve an almost 5 fold increase in iron absorption.

Enhanced Iron Absorption from Cereal and legume-Based Infant Foods by Phytic Acid Degradation

In our laboratory, we have made extensive studies on iron absorption from soy formula and infant cereals after phytic acid removal or degradation. Phytic acid was removed from soy isolates by acid/salt washing and ultrafiltration, and completely degraded by adding a phytase from *Aspergillus niger* (*10*). Phytic acid in infant cereal was similarly completely degraded by adding an exogenous phytase (*14*). Absorption from the infant foods, with different levels of phytic acid, was measured either in adults using radioisotopes or in infants using stable isotopes. The adult radioiron studies used the extrinsic tag technique with ^{59}Fe or ^{55}Fe and, in each study, 9-10 subjects were fed 4 different meals. The infant studies were made using the extrinsic tag technique with ^{57}Fe and ^{58}Fe stable isotopes. There were 10-12 infants per study and 2 different meals per infant. Fourteen days following the meal, iron absorption was calculated based on the incorporation of the isotopes into blood hemoglobin. The radioisotopes were quantified using scintillation counting and the stable isotopes by thermal ionization mass spectrometry.

Soy Formulas

Low-phytate soybean isolates were manufactured from soy flour by first water-extraction, then isoelectric coagulation of soy proteins by dilute acid. This was followed either by phytase treatment or dialysis after an acid/salt treatment. The product was finally sterilized and dried. In the first series of radioiron absorption studies (Figure 1)(*10*), the effect of phytate removal on iron absorption from soy protein isolates fed in a liquid meal with maltodextrin, corn oil and water was investigated. Iron absorption was measured relative to a formula containing an egg white control protein fed in the same subject. The egg white control meal had a mean iron absorption of between 5-9% and was designated a relative absorption of 100. Reducing the phytic acid in the soybean isolate from 1000 mg/100g to 400 mg/100 g had little or no influence on relative iron absorption, and only when it was degraded using the phytase to

120

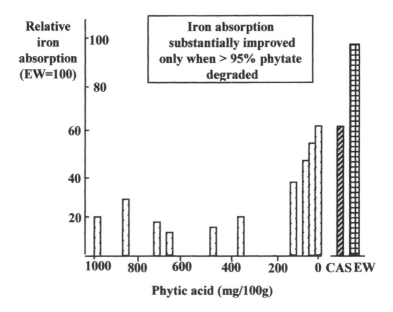

Figure 1. Effect of Phytic Acid Removal/Degradation on Iron Absorption by Adults from Soy Protein Isolates fed in a Liquid Meal Formula (10) EW= egg white; CAS = casein.

very low levels (0-30 mg/100 g) did relative iron absorption increase from around 10-20% to around 50-60% of the egg white control, the same value as given by bovine casein. Soy protein, like casein, is itself slightly inhibitory to iron absorption (*15*). The message to the food manufacturer from this study is that iron absorption from soy protein isolates can only be substantially improved when >95% of the phytate is degraded. Removing or degrading 50% or 60% is not beneficial.

In a second study, the effect of phytic acid degradation on iron absorption in adults from soy protein fractions fed in the same liquid formula meal was investigated (*15*). The major fractions of soy protein (each about 40% of the total) conglycinin (7S) and glycinin (11S) were extracted and dephytinized with phytase. Relative iron absorption from the formula containing the 7S protein was 32% and increased to 43% on dephytinization. Relative iron absorption from the formula with the 11S fraction was 21% and increased to 122% with dephytinization. The conglycinin fraction (7S) is therefore responsible for the inhibitory effect of soy protein on iron absorption, presumably due to inhibitory peptides formed on digestion. Formulas based on phytate-free glycinin (11S) would be the least inhibitory to iron absorption.

The soy isolates were next incorporated into an infant formula of similar composition to commercial infant formulas. First the effect of phytic acid removal on iron absorption in adults using the extrinsic tag radioactive isotope technique with ^{55}Fe and ^{59}Fe (16) was measured, making paired comparisons in the same subject. Iron absorption was 2.4% from a small meal of 217 g of formula containing 76 mg of phytic acid. When phytic acid was removed, this increased significantly to 6.0%. The same formulas were fed to infants and iron absorption was measured using stable isotopes (11). When phytate was 75% degraded iron absorption increased from 5.5 to 6.8% whereas, complete removal increased absorption from 3.9 to 8.7% (similar to the adult values). In both studies the differences were statistically significant (p<0.05). In a second study, the effect of ascorbic acid addition to soy formulas containing their native phytate content was investigated. The ascorbic acid content of the formula was increased from 110 mg/L in the normal commercial product to 220 mg/L, providing about 25-50 mg/meal. Under these conditions, mean iron absorption similarly increased from around 5 to 9% (11). Infant formula manufacturers can thus optimize iron absorption from soy formulas either by phytate degradation or by ascorbic acid addition, since ascorbic acid overcomes the inhibitory effect of phytic acid. Ascorbic acid converts Fe^{3+} to Fe^{2+} which binds less strongly to phytate and is also thought to bind iron in a soluble form available for absorption.

Infant Cereals

Phytic acid in infant cereals can conveniently be degraded to an undetectable level using commercial phytases. Complete degradation of phytic acid in wheat, maize, oat or rice increased iron absorption by human subjects up to 10 fold (17). However, since the level of phytate in commercial weaning foods based on low extraction flours is usually relatively low, the addition of ascorbic acid may alone be sufficient to completely remove the inhibitory effect of phytic acid on iron absorption.

When a wheat-milk infant cereal was manufactured from low-extraction wheat flour (ca. 120 mg phytic acid/100 g) and from the same wheat flour completely dephytinized, with the addition of both iron (10 mg Fe/100 g) and a generous amount of ascorbic acid (70 mg/100 g), there was no benefit to iron absorption from dephytinization (14). Iron absorption by infants was about 7.5% from both the native phytate product labeled with ^{58}Fe and the dephytinized product labeled with ^{57}Fe. It would appear that the generous amount of ascorbic acid had alone overcome the inhibition of phytate.

Ascorbic acid is relatively unstable to processing and would be rapidly oxidized during storage of weaning foods in hot and humid climates such as occur in many developing countries. An alternative to ascorbic acid is the use of EDTA compounds and these have received much attention recently (*18*) and are specifically recommended for developing countries. Many studies have demonstrated that NaFeEDTA when compared to ferrous sulfate will increase iron absorption from meals containing phytic acid (*18*). In a recent study, iron absorption from ferrous sulfate, ferrous fumarate, NaFeEDTA added to wheat, wheat-soy or quinoa infant cereals was measured in human subjects (Table I). With all iron compounds, absorption was highest from wheat cereal which contained the lowest level of phytic acid. However in all cereals, absorption from NaFeEDTA was 2-4 fold higher than from sulfate or fumarate.

Table I. Iron Absorption (%) in Adult Human Subjects from Infant Cereals Fortified with Different Iron Compounds (19)

| Infant Cereal | Fortification Compounds | | | Phytic Acid |
	Ferrous Sulfate	Ferrous Fumarate	NaFeEDTA	mg/100g
Wheat	2.2	2.1	5.2	122
Wheat-soy	0.73	0.93	2.8	770
Quinoa	0.63	1.7	1.7	763

Food Processing Methods to Remove, Degrade or Inactivate Phytate

There are 3 major methods to decrease the inhibitory effect of phytic acid on iron absorption (Table II). The first is removal, the second is enzymatic degradation and the third is to add compounds to the food which prevent phytate-mineral binding and thus inactivate the phytate. Milling of cereals can cause up to 90% reduction of phytate as it is removed in the bran together with most of the dietary fiber. Similarly dialysis or ultrafiltration of legume protein isolates after acid/salt or alkali treatment to overcome the phytate-protein binding can remove substantial quantities of phytic acid. Even more effective however, can be the soaking and germination of cereal and legume grains. These traditional processes activate the native phytases in the grains and seeds which then degrade phytic acid by successive removal of the phosphate groups. Fermentation with food grade micro-organisms similarly activates native phytases by reducing the pH, and it can also provide microbial phytases especially from yeasts and moulds. The easiest way however to degrade phytic acid is to add exogenous phytases to the cereal and legume mixtures or to adjust conditions for the maximum activity of native phytases. The commercial phytases are usually extracted from moulds, such as *Aspergillus niger*, or genetically engineered.

Table II. Major Methods to Decrease the Inhibitory Effect of Phytic Acid on Iron Absorption

Removal	- Milling of cereals (up to ca. 90% reduction) - Dialysis/Ultrafiltration of protein isolates after acid/salt or alkali treatment
Enzymatic degradation	- Soaking, germination activates native phytases - Fermentation activates native phytases, may provide microbial phytases - Steeping at optimum pH and temperature for maximum phytase activity - Add exogenous phytase
Addition of compounds which prevent phytate-iron binding	- EDTA and ascorbic acid increase absorption of iron from high phytate foods

A somewhat different approach to reducing the negative effect of phytate on iron absorption is to add other iron-binding compounds to infant foods. These compounds bind iron in the gastric juice, so preventing the reaction with phytic acid, but release it for absorption in the small intestine. EDTA compounds have been shown to positively increase iron absorption from high phytate foods (*18*) and ascorbic acid increases iron absorption by babies from soy formula (*11*) and from infant cereals fed to adult women (*20*).

Conventional Heat Treatments

Conventional heat treatments such as used in domestic cooking cause low to modest losses of phytic acid. In recent studies with mung bean or black gram (*21,22*) or maize (*23*), boiling or pressure cooking caused only 5-15% loss in the legumes, whereas with maize, popcorn had a 12% loss, boiling 18%, charcoal roasting 42% and making a chappatti 53%. The losses are presumably due to a combination of heat and/or enzyme degradation and leaching into water.

Aqueous Extraction

Aqueous extraction, especially using dilute acids, can be a reasonably effective way of removing phytic acid from legume flours. In a study by Han (24) a soybean flour containing 2.24 g phytic acid/100 g was shaken for 1h at room temperature in 30 mL water or dilute acid. Water extracted 66% of the phytic acid, 1 M sulfuric acid 68% and 1 M HCl 86%. A major disadvantage is that aqueous extraction will also remove a substantial amount of the minerals and water-soluble vitamins.

Extrusion Cooking and Roller-drying

These industrial processes may cause modest phytate losses. Extrusion cooking of rice, navy bean and a rice-soy mixture was found to reduce phytate by 20-28% although in cow-pea the reduction was 50% (25,26). In our own unpublished studies with wheat, we have found that drying an aqueous slurry of white wheat flour by extrusion cooking or roller-drying reduced the phytate by about 20%. The question remains as to whether these losses are due to enzyme degradation of phytate during some of the holding times, as seems most likely, or due to the heat process itself.

Soaking

Phytase action during soaking causes only modest phytate losses in legume seeds such as peas, groundnuts or pigeon peas (27,28). Soaking the whole seeds in excess water for 6, 12, 18 or 24 h resulted in a maximum 20% loss, with about 80% of the initial phytic acid in groundnut remaining after 24 h. This is presumably because the phytases are not in intimate contact with their substrate, the phytate, and the reaction conditions are not optimum, since studies by Gustafsson and Sandberg (29) and Sandberg (30) demonstrated that soaking finely ground material under optimal conditions can almost completely degrade phytate. Wheat bran for instance, containing 55 μmol/g phytic acid and held at 55° C in a pH 4.5-5 buffer, lost about 90% within 60 min and virtually all within 120 min. Ground brown beans containing a much lower level of phytic acid (13.6 μmol/g) held at 55° C in a pH 7 buffer lost phytic acid more slowly but lost 87% after 8 h and 97% after 18 h.

With this in mind, we recently screened a range of 26 different cereals, pseudocereals, legume and oilseed grains or seeds for the activity of their native phytases when the ground flour was held in a slurry at pH5 and 45° C (Egli et al., unpublished). Figure 2 shows selected values.

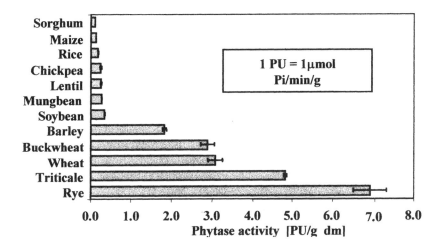

Figure 2. Activity of Native Phytase from Cereals and Legumes
(Egli et al. unpublished)

Phytase activity varied considerably. Legumes and oilseeds were generally lower than cereals or pseudocereals. Sorghum, maize and rice, however, had low phytase activity whereas rye, triticale, wheat, buckwheat and barley had the highest phytase activity. When 10% whole wheat or rye flour was added to an infant cereal containing white wheat flour and soy isolate, and the mixture held in a slurry under optimum conditions for phytase activity, phytic acid in the cereal/soy mixture was rapidly and almost completely degraded in less than one hour (Figure 3) (31).

126

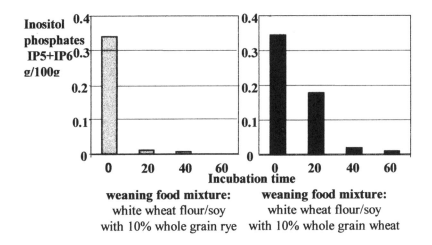

Figure 3. *Phytic Acid Degradation in Weaning Foods with Native Phytase*
(Egli et al. unpublished)

Germination

The phytic acid level of both cereal grains and legume seeds can be reduced by about 50% during germination, and a germination step can successfully be used to make low phytate weaning foods. In a study by Marero et al. (*32*), germinating rice for 3d and mung beans for 2d reduced phytic acid level by 44% and 41% respectively. After drying, dehusking, and milling, which further reduced the phytic acid, the level in the 70/30 rice-mung bean mixture used as a weaning food was only 60 mg/100g. The equivalent mixture of whole grains contained 780 mg phytic acid/100 g. Germination presumably activates the phytases, however it also stimulates other enzyme systems and both starch and protein are also degraded. A germination step would be expensive for the food industry and difficult to control, however such a pretreatment may be more useful for making traditional weaning foods in developing countries.

Fermentation

Food-grade bacteria do not usually contain phytases themselves, although they may be present in some yeasts and moulds. The benefit of fermentation in relation to phytic acid degradation is that it lowers pH through the production of organic acids and thus optimizes the conditions for the native phytases. Phytic acid in weaning foods can be completely degraded by a combination of soaking, germination and fermentation. In a study by Sharma & Kapoor (*33*), pearl millet grains which had been soaked, germinated and dried at 60° were autoclaved, made into an aqueous slurry and fermented for 4 d with *Lactobacillus acidophillus*. The original phytate content of 802 mg/100 g was reduced to 480 by soaking and to 259 by germination. No phytate was detectable after fermentation of the germinated grains.

Conclusion

Phytic acid is a major inhibitor of iron absorption in our diets and low iron bioavailability from cereal and legume-based diets in developing countries is a major cause of iron deficiency. Infants consuming cereal and legume-based weaning foods are particularly affected. Several strategies are available at both the household and the industrial level to reduce the phytic acid content of weaning foods, however, it must be virtually completely degraded in order to give a meaningful increase in iron absorption. While milling, soaking, germination and fermentation will decrease phytic acid, the most effective method is to add a commercial phytase, or a whole cereal flour rich in native phytase (rye, wheat), and to hold the weaning food mixture under optimum conditions for phytase activity until all the phytate is degraded.

References

1. Fairweather-Tait, S.J.; Hurrell, R.F. *Nutr. Res. Rev.* **1996**, *9*, 295-324.
2. Taylor, P.G.; Mendez-Castellanos, H.; Martinez-Torres, C.; Jaffe, W.; Lopez de Blanco, M.; Landaeta-Jimenez, M.; Leets, I.; Tropper, E.; Ramirez, J.; Casal, M.; Layrisse, M. *J. Nutr.* **1995**, *125*, 1860-1868.
3. Lozoff, B.; Jiminez, E.; Abraham, W.W. *New Eng. J. Med.* **1991**, *325*, 687-694.

4. *Trace elements in nutrition of children*; Chandra, R.K., Ed.; Raven Press, New York, 1987; pp.87-105.
5. Hercberg, S.; Galán, P.; Dupin, H. *Wld. Rev. Nutr. Diet.* **1987**, *54*, 201-236.
6. Reddy, N.R.; Sathe, S.K.; Salunkhe, D.K. *Adv. Food Res.* **1982**, *28*, 1-92.
7. Hallberg, L.; Rossander, L.; Skanberg, A.-B. *Am. J. Clin. Nutr.* **1987**, *45*, 988-996.
8. Brune, M.; Rossander-Hulthen, L.; Hallberg, L.; Gleerup, A.; Sandberg, A.-S. *J. Nutr.* **1992**, *122*, 442-449.
9. Larsson, M.; Rossander-Hulthen, L.; Sandström, B.; Sandberg, A.-S. *Br. J. Nutr.* **1996**, *76*, 677-688.
10. Hurrell, R.F.; Juillerat, M.A.; Reddy, M.B.; Lynch, S.R.; Dassenko, S.A.; Cook, J.D. *Am. J. Clin. Nutr.* **1992**, *56*, 573-578.
11. Davidsson, L.; Galan, P.; Kastenmayer, P.; Cherouvrier, F.; Juillerat, M.A.; Hercberg, S.; Hurrell R.F. *Pediatr. Res.* **1994**, *36*, 816-822.
12. Hallberg; L., Brune, M. and Rossander, L. *Am. J. Clin. Nutr.* **1989**, *49*, 140-144.
13. Siegenberg, D., Baynes, R.D., Bothwell, T.H., MacFarlane, B.J., Lamparelli, R.D., Car, N.G., MacPhail, A.P., Schmidt, U., Tal, A. and Mayet, F. *Am. J. Clin. Nutr.* **1991**, *53*, 537-541.
14. Davidsson, L.; Galan, P.; Cherouvrier, F.; Kastenmayer, P.; Juillerat, M.-A.; Hercberg, S.; Hurrell, R.F. *Am. J. Clin. Nutr.* **1997**, *65*, 916-920.
15. Lynch, S.R.; Dassenko, S.A.; Cook, J.D.; Juillerat, M.A.; Hurrell, R.F. *Am. J. Clin. Nutr.* **1994**, *60*, 567-72.
16. Hurrell, R.F.; Davidsson, R.M.; Kastenmayer, P.; Cook, J.D. *Br. J. Nutr.* **1998**, *79*, 31-36.
17. Hurrell, R.F.; Juillerat, M.A.; Burri, J.; Reddy, M.; Cook J.D. unpublished results.
18. International Nutritional Anemia Consultative Group (INACG)). *Iron EDTA for Food Fortification.* The Nutrition Foundation/ILSI: Washington DC.; 1993
19. Hurrell, R.F.; Reddy, M.B.; Burri, J.; Cook, J. D. *Brit. J. Nutr.* **2000**, *84*, 903-910.
20. Derman, D.P.; Bothwell, T.H.; McPhail, A.P.; Torrance, J.D.; Bezwoda, W.R.; Charlton, R.W.; Mayet, F.G.H. *Scan. J. Haematol.* **1980**, *45*, 193-201.
21. Kataria, A.; Chauhan, B.M.; Gandhi, S. *Food. Chem.* **1988**, *30*, 149-156.
22. Kataria, A.; Chauhan, B.M.; Punia, D. *Food Chem.* **1989**, *32*, 9-17.
23. Khan, N.; Zaman, R.; Elahi, M. *J. Sci. Food Agric.* **1991**, *54*, 153-156.
24. Han, Y.W. *J. Agric. Food Chem.* **1988**, *36*, 1181-1183.

25. Dublish, R. K.; Chanhan, G.S.; Bains, G.S. *J. Fd. Sci. Technol.* **1988**, *25*, 35-38.
26. Ummadi, P.; Chenoweth, W.L.; Uebersax, M.A. *J. Fd. Proc. Preserv.* **1994**, *19*, 199-131.
27. Bisnoi, S.; Khetarpaul, N.; Yadav, R.K. *Plant Food for Human Nutr.* **1994**, *45*, 381-388.
28. Igbedioh, S.O.; Kehinde, T.; Akpapunam, M.A. *Food Chem.* **1994**, *50*, 147-151.
29. Gustafsson E.L.; Sandberg, A.-S. *J. Fd. Sci.* **1995**, *60*, 149-152, 156.
30. Sandberg, A.-S. In *Nutritional and toxicological consequences of food processing* Friedman, M. Ed. Plenum Press: New York. 1991; pp. 499-508.
31. Barclay, D.; Davidsson, L.; Egli, I.; Hurrell, R.; Juillerat, M.A. Patent Application PCT/EP00/05140, publication No. WO/00/72700, 2000.
32. Marero, L.M.; Payumo, E.M.; Aguinaldo, A.R.; Matsumoto, I.; Homma, S. *Lebensm.-Wiss. u.-Technol.* **1991**, *24*, 177-181.
33. Sharma, A.; Kapoor, A.C. *Plant Foods Human* Nutr. **1996**, *49*, 241-252.

Chapter 10

Effect of Cooking on In Vitro Iron Bioavailability of Various Vegetables

Ray-Yu Yang[1], Samson T. S. Tsou[1], and Tung-Ching Lee[2]

[1]Asian Vegetable Research and Development Center, Tainan, Taiwan
[2]Department of Food Science and the Center for Advanced Food Technology, Rutgers University, 65 Dudley Road, New Brunswick, NJ 08901–8520

Iron deficiency anemia is the most prevalent nutritional problem in the world today. Diet composition is critical since low iron intake and/or bioavailability (IB) are the main causes of the deficiency. Previously, Kapanidis and Lee (*1*) showed that cooked cruciferous vegetables increased extrinsic IB three to four fold and could improve iron nutrition. The objective of the present study is to evaluate the effect of cooking on *in vitro* IB of various vegetables for their potential for improving iron nutrition and combating anemia. Forty-eight kinds of vegetables were studied for their IB (expressed as Iron dialyzability, ID) in both raw and cooked forms. A wide variation in ID was observed, ranging from 0.2% to 33.8%. Cooking resulted in higher ID in 37 of the 48 samples. Other heating processes, such as oven drying and stir-frying, were also examined and showed increased ID. We can assign these vegetables into three categories based on their ID and enhancement by cooking. Possible mechanisms for ID increases are discussed.

Introduction

More than 500 million people have iron deficiency anemia (2) and a much larger number have iron deficiency without anemia. Iron deficiency is a result of insufficient iron intake and/or low iron bioavailability in diets (3). Absorption of plant-based iron, though variable, is considered lower than that of iron from meat and is greatly influenced by interactions with enhancers and inhibitors (4, 5). Populations in developing countries with limited resources avoid hunger by consuming more plant-based food than expensive animal based products (6, 7). The population of vegetarians is increasing. Their total iron intake may meet dietary recommendations (8), however, iron deficiency still exists due to the low bioavailability of plant iron (9). Most nutrition programs aimed at decreasing iron deficiency utilize supplements and/or fortification of diets (10). An alternative and sustainable approach would be to improve the bioavailability of iron present in plant-based diets.

It was reported that cooking could enhance in vitro dialyzability of intrinsic and extrinsic iron in cruciferous vegetables previously by Kapanidis and Lee (1). The iron dialyzability of selected cruciferous vegetables, such as cabbage, kale and broccoli, can be enhanced three fold when they are boiled for a few minutes. A simple practice, such as boiling, can result in significant nutritional benefits from important micronutrients such as iron and, therefore, deserves extended research to other vegetables.

Vegetables provide multiple nutritional functions in human diets. Some are rich in micro-nutrients, some provide macro-nutrients and energy (11), while some are valued for health related factors (12). In addition to their nutrient content, vegetables are consumed in order to provide a varied and more attractive diet (13). Higher consumption rate is recommended for health maintenance and cancer prevention (14, 15), however, nutrients from only a single vegetable are insufficient for our needs. There are strong complementary nutritional effects between vegetables. Thus, it is important toassess the characteristics of induvidual vegetables. By combining vegetables with complementary characteristics, it is possible to improve the nutrition status of population using a food based approach rather than fortification.

Prolonged cooking has been reported to reduce iron absorption from meat (16). A number of studies have showm that heat can dramatically influence iron bioavailability in various foods (17) although the results are inconsistent. Vegetables vary in appearance and composition. A variety of ID changes may occur during and/or after the application of heat, since heating affects the food matrix and alters the iron chemistry (17). With the observations from cooking enhancement on the ID of raw cruciferous vegetables, further understanding the cooking effect on iron bioavailability of various vegetables would help to select specific vegetables as better iron sources and ultimately find ways to improve

132

plant based iron bioavailability. However, there is no systematic study on the IDs of cooked vegetables available in literature.

Forty-eight food commodities, mostly vegetables with a few legumes and cereals, are included in the present study. Their IDs were estimated by an *in vitro* method for comparison purposes. The objectives of this study are to evaluate the enhancing effect of cooking on *in vitro* IB of various vegetables, to assign various vegetables into separate categories based on their IDs and enhancing effects by cooking, and to provide information on their potential for improving iron nutrition and combating anemia.

Materials and Methods

Glassware and reagents

All glassware used was washed with distilled water, left overnight in a 1 M HCl bath to remove possible iron contamination and rinsed several times with distilled deionized water

Pepsin solution: 16 g pepsin 1:10,000 powder from hog stomach (Sigma) was dissolved in 100 ml 0.1 M HCl (trace metal grade, Merck). The solution was prepared fresh daily.

Pancreatine-bile (PB) suspension: 1 g of pancreatine from porcine pancreas (Sigma) and 6.5 g porcine bile extract (Sigma) were suspended in 250 ml 0.1 M $NaHCO_3$. The suspension was prepared fresh daily.

Batho reagent: 125 mg of bathophenanthrolin disulfonic acid disodium salt (Sigma) and 50 g of hydroxylamine hydrochloride (Sigma) were dissolved in 2 M sodium acetate and adjusted to a final volume of 500 ml. The solution was prepared fresh daily and prior to colorimetric determination.

Protein precipitation solution: 50 g of trichloroacetic acid and 50 ml conc. HCl were dissolved in deionized water and adjusted to a final volume of 500 ml.

Iron stock solution: 1015 ppm Fe^{+3} stock solution (Merck) was used for preparing standard solutions used in the AAS and bathopheanthroline methods.

Preparation of vegetables

Vegetables were purchased twice from a local market in Shanhua, Tainan (southern Taiwan). To avoid viscosity related errors, precisely adjusting water content with known dry matter content of samples is crucial for the relative ID comparisons among different kinds of vegetables. The samples purchased first were used for dry matter analysis only, while those purchased second were used for the determination of ID and total iron content. Approximately 2 kg of the edible portion of the fresh sample was washed with deionized water, air dried for 30 minutes, cut into 3x5 cm pieces, and well mixed.

Cooked in boiling water

An approximate 100-150g sample of vegetable was put into a 600 ml beaker. An appropriate quantity of deionized water, calculated according to the known dry matter content, was added to provide a mixture with creamy consistency with about 5% dry matter. Cooking was performed on hot plate (Thermolyne) for 15 minutes measured at the start of boiling. The mixture was stirred occasionally. Deionized water was added back to account for water loss due to evaporation. The cooked meal was quickly cooled in a water bath at room temperature (around 25°C). The vegetables, raw and cooked were ready for ID determination.

Cooked with stir-frying

Stir-frying was performed with a stainless steel wok (40 cm I.D., 9.5 cm central depth. 6.5 L capacity) on a gas stove. 40ml soybean oil was used to cook 150 g of vegetable. The oil was first heated to reach smoking point, then vegetable pieces were added and stir fried for 4 to 5 min. Finally, the cooked vegetables were transferred to a plate and cooled to room temperature (25°C) ready for ID determination.

Hot air and freezing vacuum drying

Fresh vegetables for hot air drying before cooking were cut, weighed and dried in an oven at 80°C for 16 hours. The moisture content of the dried sample was below 5-10%. Samples for freezing vacuum drying before cooking were cut and frozen to −70°C for 2 hr and dried in a freeze dryer (Virtis) for 2-3 days until the moisture content of sample decreased to below 10%.

In vitro determination of iron dialyzability

The simulated digestion method introduced by Miller and others (*18*) with minor modifications by Kapanidis and Lee (1) was used for the *in vitro* determination of iron bioavailability in this study. Raw and cooked vegetables with appropriate amounts of water were blended in a Warring blender at speed 4 for 1.5 minutes and then the pH was adjusted to 2.00±0.05 with 6 N HCl. Four aliquots of 20 g were taken in 125 ml volumetric flasks. 0.75 ml pepsin solution was added and the flasks were covered and incubated at 37°C for 2 hr. After pepsin digestion, the flasks were removed and the titratable acidity was determined using one of the aliquots. The titration stop point is at pH 7.00±0.05. 25 ml of a solution containing an amount of $NaHCO_3$ equivalent to the titratable acidity was placed in a presoaked dialysis tubing (6000-8000 m.w. cut-off number), submerged in the digests and incubated at 37°C for 30 min. 5 ml of PB (Pancreatine-bile) suspension was then added outside the tubing and continued incubation for 2 hr. After removing and rinsing, we transferred and weighed the contents. The iron concentration in the dialyzate (i.e. ID, dialyzable iron) was determined by the bathopheanthroline method (1).

Determination of available iron in dialyzate

5 ml of dialyzate was mixed with 2.5 ml protein precipitation solution in a 15 ml centrifuge tube. The tubes were placed in boiling water for 10 min, cooled, and centrifuged at 5000 rpm for 10 min. 3 ml of the supernatant was taken in another tube and 2 ml of batho reagent was added. After vortexing and standing for 15 min, the absorbance was measured at 535 nm. A calibration curve was established using 0, 0.1, 0.2, 0.3, 0.5, 1.0, and 2.0 ppm Fe standards.

The equation for calculation of ID was:

$$ID = [iron]_{dial} * Wt_{dial} / ([iron]_{veg} * Wt_{veg} / (Wt_{veg} + Wt_{water}))$$

Where $[iron]_{dial}$ = iron in dialyzate (ug/g)
            $[iron]_{veg}$ = iron in raw vegetable (ug/g)
            Wt_{dial} = Weight of dialyzate after dialysis (g);
            $Wt_{veg} + Wt_{water}$ = Weights of vegetable and water used in the preparation of initial homogenate (g)

Available iron content of the vegetable was calculated from ID multiplied by the total iron content.

Determination of iron and dry matter in vegetables

The cut vegetable was dried at 45°C for two days, and then ground to pass through a 0.5 mm sieve. One gram of powder was used for dry matter determination by heating in an over at 135°C for 2 hours. Value of dry matter was calculated after the treatment.

One gram of dried vegetable powder was weighed in a crucible for ashing and acid digestion. Ashing was performed at 600°C for 4 hours. The ashed sample was weighed and digested on a hot plate with 10 ml 3 N HCl followed by another 10 ml 1.2 N HCl, then diluted to 25-50 ml with deionized water and filtered. Iron concentration was determined using an Atomic Absorption Spectrophotometer (AAS, Z-6000, Hitachi) with an air-acetylene flame. Measurements were taken at 248.3 nm. A calibration curve was established using the Fe standards mentioned above.

Corrections between IDs and other compositions in vegetables

The SAS CORR statistical analysis procedure was used to correlate the ID of raw and cooked vegetables with their composition.

Results and Discussions

Enhancement of *in vitro* iron bioavailability of selected vegetables, legumes and cereals by cooking

A total of 48 food commodities commonly consumed in Asian diets were included in this study. They were purchased at local markets in Tainan (a city in southern Taiwan) during the summer and winter seasons of 1998. To broaden the representation of different commodities, various types of vegetables, including leaf, root, stem, flower, fungi, and fruit were selected. Legumes and cereals such as soybean, mungbean, rice and wheat were also included.

The *in vitro* iron bioavailability of these selected food commodities, with and without cooking, were determined and expressed in percentage iron dialyzability (Table I). A wide variation of iron dialyzability in both raw and cooked samples was observed. Cooking enhanced the iron dialyzability in most (37 out of 48) of the commodities tested. The ranges in percentage of iron dialyzability observed were 0.2% to 25.3% and 0.8% to 33.8%, respectively for raw and cooked samples.

Table I. Iron dialyzability of selected vegetables

Category	Vegetable	Type	Dry matter g/100 g fresh	Total iron mg/100 g dry	Iron dialyzability	
					Raw	Cooked
Category one	Amaranth	Leaf	5.6	18.5	5.1%	11.9%
	Asparagus	Stem	6.2	6.8	5.6%	16.0%
	Bamboo shoot	Stem	7.1	9.3	4.6%	10.2%
	Borecole	Leaf	6.3	10.6	3.5%	20.3%
	Broccoli	Flower	9.6	6.7	6.0%	31.3%
	Cabbage	Leaf	5.3	6.5	4.2%	24.5%
	Cauliflower	Flower	7.1	7.2	6.7%	26.7%
	Coriander	Leaf	6.2	31.7	1.3%	8.1%
	Corn, fresh	Cereal	21.1	2.6	12.3%	21.3%
	Kidney bean	Legume	7.7	8.3	2.9%	13.7%
	Leek flower	Onion	7.2	8.3	5.2%	16.1%
	Lima bean	Legume	31.3	6.9	6.6%	11.9%
	Onion fragrant	Onion	6.2	9.6	7.5%	16.1%
	Paitsai	Leaf	3.2	18.3	2.1%	11.0%
	Pumpkin	Fruit	9.6	4.3	8.7%	21.5%
	Radish	Root	6.4	7.7	0.6%	9.0%
	Wax-gourd	Fruit	2.4	6.9	2.5%	11.5%
Category two	Basil	Leaf	10.4	19.9	0.7%	2.7%
	Carrot	Root	13.7	8.6	4.6%	7.7%
	Celery	Leaf	4.2	22.0	0.4%	1.9%
	Cucumber	Fruit	2.8	9.4	0.5%	3.3%
	Dasheen	Stem	32.6	4.9	5.1%	5.1%
	Eggplant	Fruit	6.1	6.1	2.4%	1.4%
	Garland chrysanthemum	Leaf	3.4	26.8	1.7%	2.4%
	Indian bean	Legume	86.0	6.5	5.4%	3.3%
	Kang-kong	Leaf	4.9	12.0	4.5%	4.3%
	Lettuce	Leaf	3.3	11.4	1.5%	3.1%
	Lotus	Root	16.0	4.3	8.6%	8.2%
	Mungbean	Legume	88.0	6.0	3.6%	5.0%
	Mushroom	Fungi	8.5	12.2	6.9%	4.0%
	Mustard leaves	Leaf	3.8	33.8	1.3%	5.3%
	Onion	Stem	10.7	2.9	6.7%	7.2%
	Pea with pod	Legume	9.9	10.6	6.7%	9.9%
	Rag gourd	Fruit	4.6	6.2	7.8%	7.1%

Table I. *Continued*

Category	Vegetable	Type	Dry matter g/100 g fresh	Total iron mg/100 g dry	Iron dialyzability Raw	Iron dialyzability Cooked
	Red leaves	Leaf	5.9	18.2	0.2%	3.4%
	Rice, milled	Cereal	84.5	1.4	0.7%	1.1%
	Soy bean	Legume	88.2	8.4	2.5%	5.4%
	Spinach	Leaf	6.0	39.8	7.2%	7.2%
	Vegetable Soybean	Legume	33.0	8.8	3.2%	7.7%
	Wheat, whole	Cereal	88.0	4.6	0.3%	0.8%
Category three	Bitter melon	Fruit	5.0	7.0	10.1%	14.6%
	Ginger	Stem	4.7	11.7	25.3%	20.9%
	Pepper, green	Fruit	5.5	5.4	16.7%	32.4%
	Pepper, red	Fruit	13.4	5.8	23.7%	29.1%
	Sweet potato	Root	31.3	2.4	15.2%	14.8%
	Tomato	Fruit	4.8	6.0	24.6%	33.8%

Based on the iron dialyzability of raw samples and the cooking effect, one may divide these commodities into three categories:

1) Low ID in raw samples. The ID can be substantially enhanced (more than 2 fold) through the process of cooking. The ID after cooking is increased to about 10% or more. 35% of samples tested belong to this category.

2) Low ID in raw samples. Cooking does not enhance ID substantially. 50% of the total samples tested fall into this group. The dialyzable iron is below 10% of total iron after cooking.

3) ID is above 10% in raw samples. Cooking does not have significant enhancing effect for most samples belonging to this group.

As shown in Table 1, there is no consistent trend between cooking effect and the type of vegetable. Different types of vegetable (leaf, root, stem, and fruit) are found in both categories one and two. No leafy vegetables were found in category three in this study.

Available iron in crops varies from 5 ug/100g for eggplant to 397 ug/100g for soybean. In spite of low iron bioavailability, most legumes, such as mungbean, vegetable soybean, lima bean and pea with pod, are high in available iron due to their higher iron content. This suggests that legumes deserve more attention as iron sources especially in regions where pulses and beans are consumed regularly in relatively high quantities.

Vegetables in category three, such as ginger, tomato and red pepper, are rich iron sources and can be consumed either as a salad or cooked with other ingredients. However, vegetables in category one can be a good source of iron only after cooking.

The two tested cereals (milled rice and whole wheat) were both low in iron contents and ID. They belong to category two as cooking does not enhance their ID. However, they play a significant role in most diets as a source of iron due to their high level of consumption.

Enhancing effect of other heating processes on ID

Cabbage was used to study the enhancing effect of oven drying and stir-frying on ID. The results are summarized in Figure 1. Stir-frying with soybean oil is a common practice for preparing vegetables in Asia. Its effect on ID was found to be comparable to that of boiling. Oven drying, however, is less effective at enhancing iron availability. Moreover, cooking dried cabbage did not further enhance the ID. To study the effect of drying, freeze-drying was applied to dehydrate cabbage samples. Further cooking was not able to recover the availability of iron as compared with cooked fresh samples (Figure 1). Iron in cabbage became less available after drying and cannot be released as the amount of that through cooking. Iron also became less available as a result of refrigeration for a prolonged period. Figure 2 shows a clear decrease in ID with increased storage period.

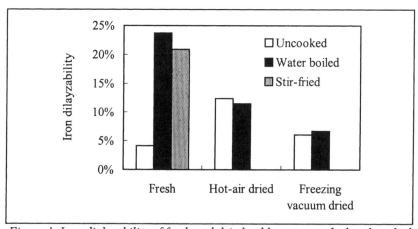

Figure 1. Iron dialyzability of fresh and dried cabbage, uncooked and cooked

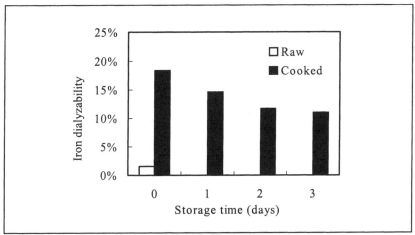

Figure 2. Effect of cold storage at 4 ℃ on iron dialyzability of cooked cabbage

Correlations between IDs and other compositions in vegetables

It is known that iron bioavailability of vegetables is strongly affected by many factors (*19*). Meat (*20*) and ascorbic acid (*21, 22, 23*) increase iron bioavailability whereas fiber (*24*), phytate (*25*), polyphenols (*26, 22*), calcium (*27*), and certain proteins (*28*) act as inhibitors. The physical and chemical properties of these constituents and the enzymes associated with them may be changed in the heating process. Our study attempted to correlate the possible roles of some major constituents of raw and cooked vegetables to their involvement in cooking enhancement of ID. The SAS CORR procedure for statistical analysis was used to correlate the ID of raw and cooked vegetables with their composition. The values of constituent contents were taken from the food composition table of Taiwan (*11*). The composition values included in the analysis were protein, fat, carbohydrate, iron, phosphorous, vitamins A, B_1, B_2, and C, and niacin. Only vegetable samples with dry matter content lower than 50% were applied to this analysis, and thus, the number of samples selected was 43. ID of raw vegetables was significantly negatively correlated to protein content (r=-0.327*) and calcium content (r=-0.311*), and positively correlated to carbohydrate content (r=0.371*) at p=0.05, whereas the ID of cooked vegetables was positively correlated only to vitamin C content (r=0.344*) at p=0.05. The result coincided with the observations: protein and calcium are inhibitors of iron bioavailability (*28, 27*). The inhibitory effect of protein and calcium was reduced by cooking. The correlation coefficients between the ID of cooked vegetables against protein and calcium were not significant, r = -0.110

and r = -0.118, respectively. A reduction of positive correlation between the ID and carbohydrate content was also observed after the vegetable samples were cooked (r = 0.110). The reduction of inhibitory effect of protein and calcium by cooking could be due to changes in the matrix of the tissue.

It was noted that the correlation coefficient between the ID of uncooked vegetables and vitamin C is relatively low (r =0.099) though it has been regarded as an enhancer of iron absorption. With its reducing power, it renders iron in the form of Fe^{+2} at pH 2.6-6.0 (29). Formation of an iron–ascorbic acid complex is also reported to be able to maintain iron in a soluble form in the intestine with a neutral pot (30). In this study, the relatively low correlation of vitamin C content with ID of uncooked vegetables and significant correlation obtained after vegetables were cooked suggests that the enhancing effect of vitamin C can be more efficient once inhibitory factors are destroyed by cooking.

Conclusion

The primary objectives of this study were to generate information on iron bioavailability of vegetables in order to help guide food-based approaches to alleviate dietary iron deficiency, and to better understand the possible mechanisms of the effect of cooking on iron bioavailability for their potential for improving iron nutrition and combating anemia. From the observations based on our present study, we may conclude that:

•   The enhancing effect of cooking on the iron dialyzability of vegetables is confirmed in an increased number of commodities. This effect, however, is not observed in cereals, legumes, and certain vegetables. Information in categorizing vegetables based on their iron dialyzability in response to cooking is important for practical use. Through selection of vegetables consumers will be able to obtain more available iron without too much change in their dietary patterns.

•   An enhancing effect can be achieved with different heating processes, such as boiling, stir-frying and even hot air drying. It is better to consume the vegetable dishes freshly prepared. Prolonged storage of cooked vegetables in refrigeration will reduce the availability of iron.

•   It is speculated that the removal of heat sensitive inhibitors is involved as part of the mechanism of the enhancing effect on iron bioavailability. Further studies will be required. Cabbage and tomato, representing categories 1 and 2, will be good model crops for future studies.

•   Most of the vegetables have relatively low iron content. Approaches of enhancing the iron dialyzability of legumes should be considered in order to provide sufficient available iron in diets of vegetarians and populations who are not able to consume sufficient meat.

Acknowledgements

This study was supported by a grant from the USAID OMNI Research Program through the Human Nutrition Institute of the International Life Sciences Institute (ILSI). The opinions expressed herein are those of the authors and do not necessarily represent the views of ILSI. We would also like to thank Professor Ning-Sing Shaw of National Taiwan University, Taiwan for her communication.

References

1. Kapanidis, A.N.; Lee T-C. *J. Food Sci.* **1995**, *60*, 128-131, 141.
2. *Third Report on the World Nutrition Situation*; ACC/SCN: Geneva, Switzerland, 1997; pp 34-40
3. Fairweather-Tait, S.J. *J Royal Soc. of Health.* **1984**, *104*, 74-78.
4. Layrisse, M.; Cook, J.D.; Martinez, C.; Roche, M.; Kuhn, I.N.; Walker, R.B.; Finch, C.A. *Blood* **1969**, *33*, 430-443.
5. Cook, J.D. *Food Technology* **1983**, *37*, 124-126.
6. Baker, S.J.; deMaeyer, E.M. *Am. J. Clin. Nutr.* **1979**, *32*, 368-417.
7. *Enriching Lives. Overcoming vitamin and mineral malnutrition in developing countries.* World Bank: Washington DC, 1994.
8. *Requirements of vitamin A, iron, folate and vitamin B12: Report of a Joint FAO/WHO consultation*; FAO/WHO: Rome, Italy, 1988; pp 33-50.
9. Craig, W. J. *Am. J. Clin. Nutr.* **1994**, *59*, 1233S-1237S.
10. Scrimshaw N.S. *Food and Nutrition Bulletin* **1996**, *17*, 1-41
11. FIRDI/PUS *Food composition table in Taiwan Area;* Public Health Division, Executive Yuan: Taipei, Taiwan, 1998; pp 100-108.
12. *Phytochemical Dictionary – a Handbook of Bioactive Compounds From Plants*, 2nd ed.; Harborne, J.B.; Baxter, H.; Moss, G.P., Eds; Taylor and Francis Ltd.: Philadelphia, PA, 1999.
13. Willett, W.C. *Science (Washington DC)* **1994**, *264*, 532-537.
14. Steinmetz, K.A. *J. Am. Diet. Assoc.* **1996**, *96*, 1207-1039.
15. Law M.R.; Morris J.K. *Eur. J. Clin. Nutr.* **1998**, *52*, 549-556.
16. Martinez-Torres, C.; Leets, I.; Taylor, P.; Ramirez, J; del Valle Camacho, M.; Larisse, M. *J. Nutr.* **1986**, *116*, 1720-1725
17. Lee, K. In *Nutritional Bioavailability of Iron;* Kies, C., Ed.; American Chemical Society: Washington DC, 1982; pp 27-54.

142

18. Miller, D.D.; Schricker, B.R.; Rasmussen, R.R.; Van Campen, D. *Am. J. Clin. Nutr.* **1981**, *34*, 2248-2256.
19. Monsen, E.R. *J. Am. Diet. Assoc.* **1988**, *88*, 786-790.
20. Mulvihill, B.; Morrissey, P.A. *Food Chem.* **1998**, *61*, 1-7.
21. Hoffman, K.E.; Yanelli, K.; Bridges, K.R. *Am. J. Clin. Nutr.* **1991**, *54*, 1188-1192.
22. Siegenberg, D.; Baynes, R.D.; Bothwell, T.H.; MacFarlane, B.J.; Lamparelli, R.D.; Car, N.G.; MacPhail, P.; Schmidt, U.; Tal, A.; Mayet, F. *Am. J. Clin. Nutr.* **1991**, *53*, 537-541.
23. Hazell, T; Johnson, I.T. *Brit. J. Nutr.* **1987**, *57*, 223-233.
24. Cook, J.D.; Nobel, N.L.; Morck, T.A.; Lynch, S.R.; Petersburg, S.J. *Gastroenterology* **1983**, *85*, 1354-1358.
25. Torr, M.; Rodriguez, A.R.; Saura Calixto, F. *CRC Crit. Rev. Food Sci. Nutr.* **1991**, *1*, 1-22.
26. Disler, P.B.; Lynch, S.R.; Charlton, R.W.; Torrance, J.D.; Bothwell, T.H. *Gut* **1975**, *16*, 193-198.
27. Hallberg, L.; Brune, M.; Erlandsson, M.; Sandberg, A.S.; Rossander Hulten, L. *Am. J. Clin. Nutr.* **1991**, *53*, 112-119.
28. Kratzer, H.F.; Vohra, P. In *Chelates in Nutrition;* CRC: Boca Raton, FL, 1986; pp 97-114.
29. Hsieh, Y-H. P.; Hsieh, Y.P. *J. Agric. Food Chem.* **1997**, *45*, 1126-1129.
30. Gorman, J.E.; Clydesdale, F.M. *J Food Sci.* **1983**, *48*, 1217-1220,1225.

Chapter 11

Stability of Biologically Active Pyridoxal and Pyridoxal Phosphate in the Presence of Lysine

Tzou-Chi Huang[1], Ming-Hung Chen[1], and Chi-Tang Ho[2]

[1]Department of Food Science, National Pingtung University of Science and Technology, 912, Pingtung, Taiwan
[2]Department of Food Science, Rutgers University, 65 Dudley Road, New Brunswick, NJ 08901–8520

Stability of the biologically active compound, vitamine B6 in aqueous solution, was investigated. Schiff base formation is the major reaction between the ε–amino group of lysine and the aldehyde group of both pyridoxal and pyridoxal phosphate. Model systems composed of an equal molar of lysine with either pyridoxal or pyridoxal phosphate were used to study the effect of proton transfer on Schiff base formation. Pyridoxallysine was found to be the major product in both lysine/pyridoxal and lysine/pyridoxal phosphate systems. Quantitation of residual pyridoxal and pyridoxal phophate was conducted to evaluate the degradation of pyridoxal and pyridoxal phophate. The results indicate both the free phosphate ion in the buffer system and the bound phosphate on pyridoxal phosphate can enhance the formation of the Schiff base. The phosphate group serves as both proton donor and acceptor, which catalyzes the Schiff base formation. The aldehyde group on pyridoxal phosphate was found to be much more reactive than that on pyridoxal. The bound phosphate group on pyridoxal phosphate, with proton donating and accepting groups in close proximity, can simultaneously donate and accept protons, thus enhancing the Schiff base formation between the aldehyde group and the ε-amino group. The deterioration rate of pyridoxal phosphate was faster than that of pyridoxal in an aqueous system.

Introduction

Considerable research has focused on the bioavailability of vitamin B6 and lysine from foods. The thermal degradation of lysine has been extensively studied (*1*), whereas less research has been conducted to establish mechanisms for vitamin B6 degradation (*2-3*).

Cereals may supply nearly 60-70% of the protein in diets. Wheat has higher protein among cereals but lysine is still the limiting amino acid in wheat protein. Several studies showed that fortification of white flour with synthetic lysine markedly improves the nutritive value of wheat proteins (*4*). Post-processing fortification of wheat flakes with 0.28% lysine significantly improved protein quality parameters, protein efficiency ratio (PER) and net protein utilization (NPU) of the product compared to the control (*5*). Increased PER values were obtained by adding 0.3-0.5% L-lysine HCl or by combining peanut with soy proteins (*6*).

It is usual to calculate vitamin B6 requirements relative to protein intake. Current RDAs range between 1.5 and 2.2 mg/d. A minimum safe intake is 11 μg/g dietary protein. Higher intakes are required in pregnancy and lactation, and possibly also in the elderly. Average intakes of vitamin B6 in developed countries meet the target of 15 μg/g dietary protein. Biochemical evidence showed an inadequate vitamin B6 nutritional status in 10-25% of the population. To improve commonly consumed cereal grain products, which are prime suppliers of calories and protein, the Food and Nutrition Board of the National Research Council has recommended that all products based on wheat, corn, and rice have the following nutrients added: vitamin B6, vitamin A, thiamin, riboflavin, niacin, folacin, iron, calcium, magnesium, and zinc (*7*).

Varying extent of losses of vitamin B6 resulting from processing and preservation of foods were studied. Main losses of vitamin B6 were reported to vary from 57.1 to 77.4 % in canned vegetables to 42.6 to 48.9 % in canned fish, seafood, meat and poultry (*8*). Conflicting data have been reported concerning the effects of processing on the bioavailability of naturally occurring vitamine B6 in food products.

The five major forms of vitamin B-6 are pyridoxal 5'-phosphate (PLP), pyridoxal (PL), pyridoxamine (PM), pyridoxamine 5'-phosphate (PMP) and pyridoxine (PN).

Among them, pyridoxal 5'-phosphate is an active coenzyme in many metabolic transformations of amino acids and may play a role in their absorption and transport. Plasma pyridoxal phosphate concentrations less than 34.4 nM were observed in 26% of girls and values from 34.4 to 40.5 nM, in 14% (9). The effect of aging upon vitamin B-6 status has been examined (10). The results of a study of plasma pyridoxal 5'-phosphate levels in various age groups (3 hours old to over 80 years old) shows a gradual decrease in plasma PLP from 137 nmol/L to 23.1 nmol/L. Blood plasma was collected from approximately 2500 elderly subjects born between 1913-1918 living in 17 small towns in 11 European countries, and the plasma levels of carotene, retinol, alpha-tocopherol, vitamin B_{12}, folic acid and pyridoxal 5'-phosphate were determined (11). The vitamin status for retinol and folic acid was adequate in all countries, whereas the prevalence of biochemical vitamin B_6 deficiency was widespread and reached over 50% in some countries.

The bioavailability of vitamin B6 in foods is believed to be affected by the thermal formation of ε -pyridoxyllysine (12). However, the effect of buffer on vitamin B6 bioavailability has not been fully established. Gregory and Hiner (13) revealed that at pH 7, pyridoxal phosphate was 1.5-2.0 fold less stable than pyridoxal. However, opposite results were obtained by the same research group (3). They found that PLP exhibited slightly greater stability than PL towards caseinate under the reaction condition of autoclaving at 121°C for 45 min.

The *in vitro* synthesis of 3-carboxy-1-pyridoxyl-1,2,3,4-tetrahydro-β-carboline and 3-carboxy-1-pyridoxyl-1,2,3,4-tetrahydro-β-carboline 5'-phosphate from tryptophan with PL and PLP both 0.1 mM respectively, under physilogical conditions, pH 7.0 at 37°C is presented by Argoudelis (14). Interestingly, *in vivo* formation of similar cyclic compounds was reported from L-dopa and PLP or PL in the liver and blood of rats (15) and from histamine and PLP or PL in the tissues and urine of mice (16) when large quantities of L-dopa or histamine were given to the animal. Although, plasma PLP was as low as 45±2 nmol/L initially, it reached 377±12 nmol/L after 7 days of supplementation (17). The possibility for the interaction between L-lysine and PLP to proceed *in vivo* increases as the concentration of L-lysine increases in blood when excess L-lysine is given the animals or in the lysine-fortified cereal foods. In this paper, the *in vitro* synthesis of pyridoxyllysine under physiological conditions, pH 7.2 and 37°C are presented. This information will be valuable to investigators studying the bioavailability and reactivity of PL and PLP toward amino acids, especially lysine, in drug, in animals or humans, and the possible interaction of certain pyridoxal 5'-phosphate requiring enzymes.

Experimental Procedure

Reaction mixtures were composed of equal molar solutions of pyridoxal or pyridoxal phosphate and L-lysine (0.1 M) in deionized water, adjusted to pH

7.2 with sodium hydroxide. To study the effect of the buffer capacity on pyridoxyllysine formation, solutions containing pyridoxal (0.1 M) and L-lysine (0.1 M) with different proportions of anion phosphate (0.1 M and 0.2 M) at an initial pH of 7.2 were prepared. Each reaction mixture was routinely analyzed by HPLC with photodiode array detector with a Shimpak (C-18) column under isocratic conditions at ambient temperature. Remaining pyridoxal and pyridoxal phosphate and also pyridoxylidenelysine were detected spectrophotometrically at 295 nm and quantified against an internal standard (pyridoxine) by following the method of (13). FAB MASS analysis was utilized to confirm the formation of pyridoxylidenelysine.

Interaction of Pyridoxal and Lysine

A typical HPLC profile of a heated pyridoxal/lysine mixture is shown in Figure 1. Peak 2 was characterized as pyridoxal by spiking with an authentic standard. To confirm the structure of the peak 1 eluted by HPLC analysis, one of the reacted mixtures was subjected to HPLC with photodiode array used as the detector. A characteristic shoulder of pyridoxyl amino absorption maximum at 334 nm was observed in the Schiff base, indicating that a conjugated double bond formed after the coupling of the carbonyl moiety of pyridoxal molecule onto the lysine anino group. In addition to the evolution of the shoulder in the absorption spectrum of the Schiff base, the absorption maximum shifted from 295 nm to 287 nm when they exited in the mobile phase used in HPLC analysis (Figure 2). The binding of pyridoxal to lysine was confirmed by a FAB-MS spectrum. The molecular weight of this compound was found to be 295. Peak 1 was tentatively characterized as pyridoxylidenelysine. Similar pyridoxylidene-amino acids, pyridoxylideneglutamic acid, pyridoxylidenealanine (18), N6-(P-pyridoxylidene)-aminocaproic acid (19) and pyridoxylidenevaline (20), have been synthesized.

Effect of Phosphate Buffer on Pyridoxylidenelysine Formation

Concentration of pyridoxal decreases with increasing the reaction time, whereas that of peak 1 increases with increasing reaction time (Figure 3). As expected, addition of phosphate ion into the aqueous model system of

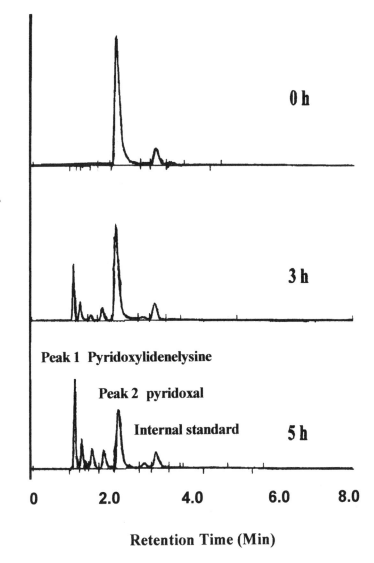

Figure 1. Typical HPLC profile of heated PLP/lysine system in aqueous solution at pH 7.2.and 37°C.

148

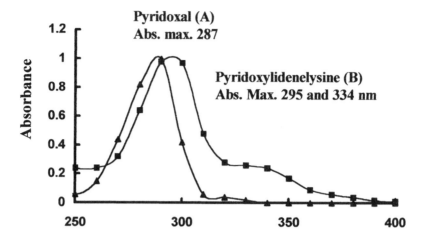

Figure 2. Absorption spectra of pyridoxal (A) and pyridoxylidenelysine (B)

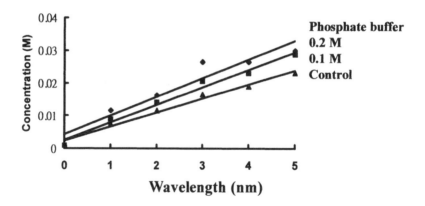

Figure 3. Effect of phosphate on pyridoxylidenelysine formation in aqueous pyridoxal-lysine system with various concentration of phosphate buffer at pH 7.2.

pyridoxal/lysine increased the Schiff base formation rate significantly, as shown in Figure 4. The pyridoxylidenelysine formation increased with increasing the concentration of phosphate ion from 0.1 M to 0.2 M. The free phosphate may catalyze the Schiff base formation between the ε-amino group of lysine and the aldehyde group of pyridoxal. Similar buffer effect has been observed in catalyzing the Schiff base formation in thiazolidine from aldehydes and cysteamine (21) and diketopeperazine from aspartame (22).

Interaction of Pyridoxal Phosphate and Lysine

A typical three dimensional HPLC profile of a heated pyridoxal phosphate/lysine mixture is shown in Figure 5. Peaks 1 and 2 were characterized as pyridoxylidenelysine and pyridoxal respectively. When the reaction mixture (37°C, 1 hr) was spiked with authentic standard, Peak 3 was characterized as pyridoxal phosphate. The data revealed that pyridoxal phosphate is an effective catalyst for PLP deterioration. As shown in Figure 6, pyridoxal phosphate was much more unstable than pyridoxal in aqueous solution in the presence of lysine at pH 7.2. Within 1 hour, all the pyridoxal phosphate were converted to pyridoxal. Pyridoxal phosphate in the system was bound to lysine through Schiff base linkage, followed by a depletion of the phosphate group, leading to the formation of pyridoxal. The interaction between pyridoxal and lysine in the system leads t o the formation of pyridoxylidenelysine.

The bound phosphate on pyridoxal phosphate could enhance the Schiff base formation between the aldehyde group of pyridoxal phosphate and the amino group of lysine. A higher yield in the synthesis of 3-carboxy-1-pyridoxyl-1,2,3,4-tetrahydro-β-carboline 5'-phosphate compared to that of 3-carboxy-1-pyridoxyl-1,2,3,4-tetrahydro-β-carboline 5'-phosphate was observed, indicating that cyclic compounds (1-tetrahydrocarbolines) are formed from L-tryptophan and pyridoxal or pyridoxal 5'-phosphate. Both data reveal that the aldehyde group on PLP was much more reactive than that on PL. Gregory III and Kirk (23) reported that PLP was much more reactive than PL. Quantitative results obtained from using a PLP/peptide system were compared with those obtained from using a PL/peptide system. The results showed that the yield was much higher for peptide-bound PLP at 20% compared to a value of 3% for peptide-bound PL.

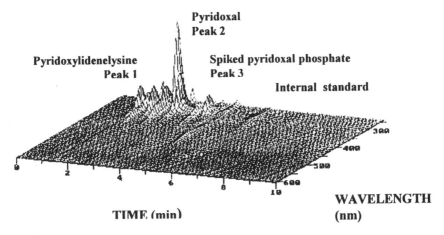

Figure 4. Three-dimensional photodiode array chromatogram of of pyridoxal phosphate/lysine model system incubated at pH 7.2 , 37°C for 1 hr.

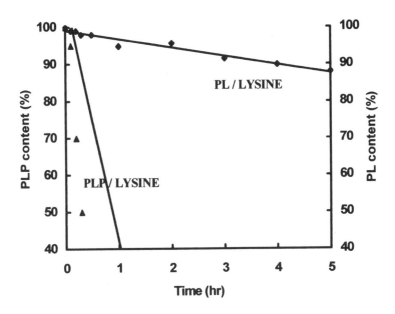

Figure 5. Stability of pyridoxal (PL) and pyridoxal phosphate (PLP) in aqueous solution at pH 7.2 (37 °C).

Kinetic Aalysis of the Degradation of Pyridoxal and Pyridoxal Phosphate

The degradation of pyridoxal increased with increasing reaction time and temperature. To calculate the Arrhenius activation energy (Ea), we regressed the remaining pyridoxal concentration with time at a constant temperature to determine the rate constant, and then, regressed ln K with the reciprocal temperature to determine Ea. The degradation rates for pyridoxal in the presence of lysine at 40°C, 50°C and 60°C were 0.0001, 0.0005 and 0.0011 M/hr respectively. A similar tendency was observed for pyridoxal phosphate degradation in the lysine/pyridoxal phosphate aqueous model system. The degradation rates for pyridoxal phosphate in the presence of lysine at 30°C, 40°C and 50°C were 0.0003, 0.0011 and 0.0017 M/hr respectively. The activation energy values were calculated from the Arrhenius plots for pyridoxal at pH 7.2 in aqueous solution, 0.1 M phosphate buffer, and 0.2 M phosphate buffer, as well as pyridoxal phosphate in aqueous solution of 23.9, 12.4, 10.5 and 5.7 Kcal/mol respectively as listed in Table 1. The presence of a phosphate ion decreased the activation energy for pyridoxal degradation. The activation energy for pyridoxal degradation also decreased with increasing the phosphate buffer concentration from 0 to 0.2 M, whereas, the activation energy values for pyridoxal phosphate degradation was the lowest, even in the absence of free phosphate ions in the reaction mixture. Kinetic analyses showed a difference in temperature dependence on the relative degradation rate of the B6 vitamers. The activation energy values for the loss of pyridoxine, pyridoxamine and pyridoxal during processing were 27.3, 23.7 and 20.8 kcal/mole respectively (13). Relatively high activation energy values for the loss of PN, PL, and PM during processing in neutral phosphate buffer were reported to be 54, 50 and 85 kcal/mole respectively (8).

Table 1. Statistical evaluation of activation energies for pyridoxal and pyridoxal phosphate degradation in the presence of lysine system at pH 7.2

Samples	Arrhenius kinetic values	
	Ea (Kcal/mol)	R^2
Pyridoxal in aqueous solution	23.9	0.95
Pyridoxal + 0.1 M phosphate buffer	12.4	0.98
Pyridoxal + 0.1 M phosphate buffer	10.5	0.91
Pyridoxal phosphate in aqueous solution	5.7	0.73

Proposed Mechanism for Pyridoxylidenelysine Formation

The interaction of PL and PLP with free amino group of amino acids, amines, peptides and proteins has been extensively studied. That reactivity of PLP was greater than that of PL. It was attributed to the blocking of internal hemiacetal formation by the bound phosphate moiety (24). Argoudelis (14) explained the higher reactivity of aldehyde in PLP than that of PL as being due to the fact that PLP exists in a free aldehyde form rather than a hemiacetal. Based on kinetic analysis and the series studies on phosphate-mediated catalysis of Schiff base formation (21,25-26), a mechanism for pyridoxylidenelysine formation was proposed, as shown in Figure 6. Pyridoxal may react with the ε-amino group of lysine to form a Schiff base, and subsequently rearrange to pyridoxylidenelysine, which was catalyzed by the nearby phosphate group. The phosphate group acts both as a proton donor and acceptor. The bound phosphate group on pyridoxal phosphate was found to be much more reactive than the free phosphate in the buffer system. The bound phosphate group on pyridoxal phosphate, being both proton donating and accepting groups in close proximity, can simultaneously donate and accept protons, thus enhancing the Schiff base formation between the aldehyde group and the ε-amino group. A similar reaction of α-amino group on lysine was observed in the study of glucosylysine formation from the Maillard reaction of glucose and lysine ε-amino group (27).

References

1. Erbersdobler, H.F.; Hartkopf, J.; Kayser, H.; Ruttkat, A. In *Chemical Markers forProcessed and Stored Foods*; Lee, T.C.; Kim H.J., Eds.; ACS Symp. Ser. 631; American Chemical Society: Washington, DC, 1995; pp 45-53.
2. Gregory III, J. F.; Ink, S. L.; Sartain, D. B. *J. Food Sci.* **1982,** *47*, 1512-1518.
3. Gregory III, J. F.; Ink, S.L.; Sartain, D. B. *J. Food Sci.* **1986**, *51*, 1345-1351.
4. Geervani, P.; Devi, P. Y. *Nutr. Rep. Internat.* **1986**, 33, 961-966.
5. Ewaidah, E. E.; Al-Kahtani, H. A. *Cereal Foods World.* **1992**, *37*, 386-388, 390-391.
6. Bookwalter, G. N.; Warner, K.; Anderson, R. A.; Bagley, E. B. *J. Food Sci.* **1979**, *44*, 820-825.
7. Christopher, B. J. *Hunger Notes - World Hunger Education Service 8*, 6.

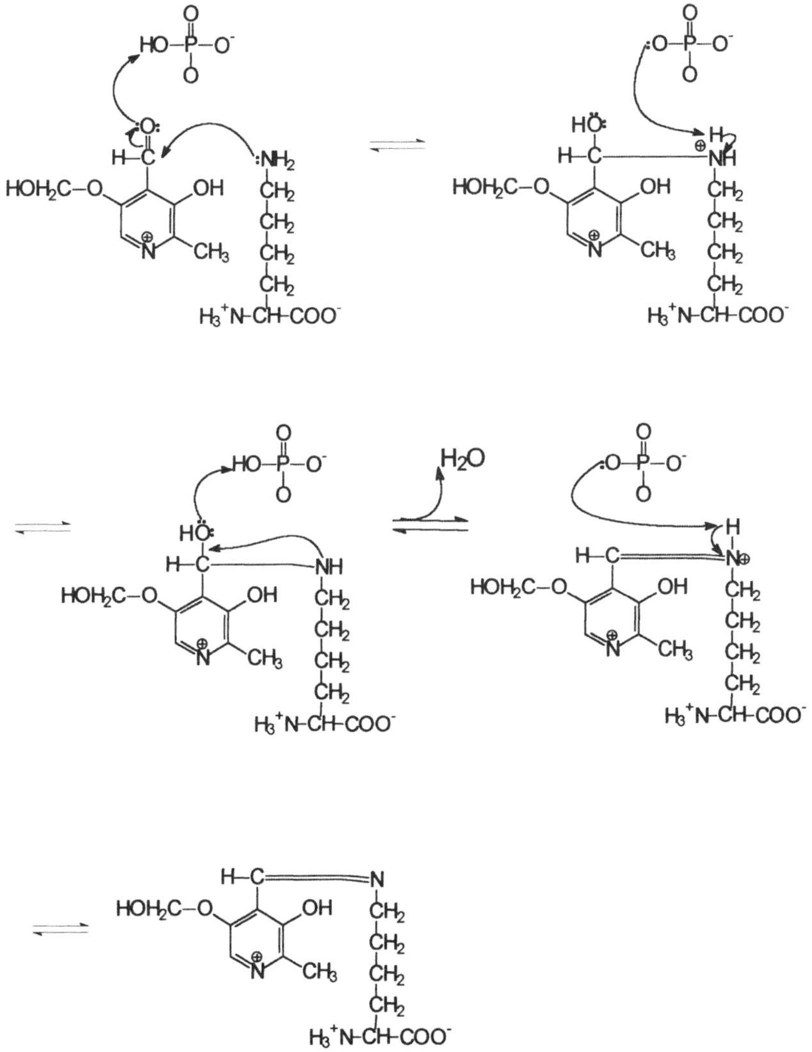

Figure 6. Proposed formation mechanism for pyridoxylidenelysine phosphate from pyridoxal and lysine.

154

8. Navankasattusas, S.; Lund, D. B. *J. Food Sci.* **1982**, *47*, 1512-1518.
9. Driskell, J. A.; Moak, S. W. *Amer. J. Clin. Nutr.* **1986**, *43* , 599-603
10. Hamfelt, A.; Soderhjelm, L. *Current Topics in Nutrition & Disease.* **1988**. *19*, 95-107.
11. Haller, J.; Lowik, M. R. H.; Ferry, M.; Ferro-Luzzi, A. *Euro. J. Clin. Nutri.* **1991**, *45*, 63-82.
12. Gregory, J. F.; Kirk, J. R. *J. Food Sci.* **1978**, *43*, 1801-1815.
13. Gregory III, J. F.; Hiner, M. E. *J. Food Sci.* **1983**, *48*, 1323-1326.
14. Argoudelis, C. J. *J. Agric. Food Chem.* **1994**, *42*, 2372-2375.
15. Bringmann, G.; Schneider, S. *Angew. Chem.*, Int. Ed. Engl. **1986**, *25*, 177-178
16. Kierska, D.; Sasiak, K.; Maslinski, C. *Agents Actions.* **1981**, *11*, 28-32.
17. Kang-Yoon, S. A. Kirksey, A. *Amer. J. Clini. Nutri.* **1992**, *55*, 865-872.
18. Yoshihiko, M. *J. Amer. Chem. Soc.* **1957**, *79*, 2016-2019
19. Schonbeck,.N. D.; Skalski, M.; Schfer, J. A. *J. Biol. Chem.* **1975**, *250*, 5343-5351
20. Metzler, C. M.; Cahill, A. Metzler, D. E. *J. Amer. Chem. Soc.* **1980**, *102*, 6075-6082
21. Huang, T. C. Huang, L. Z. Ho, C.-T. *J. Agric. Food Chem.* **1998**, *46*, 224-227.
22. Bell, L. N.; Wetzel, C. R. *J. Agric. Food Chem.* **1995**, *43*, 2608-2612.
23. Gregory III, J. F.; Kirk, J. R. *J. Food Sci.* **1977**, *42*, 1354-1351.
24. Heyl, D.; Luz, K.;Harris, S. A.; Folkers, K. *J. Amer. Chem. Soc.* **1951**, *73*, 3430-3433
25. Huang, T. C.; Fu, H. Y.; Ho, C.-T. *J. Agric. Food Chem.* **1996**, *44*, 240-246.
26. Huang, T. C.; Su, Y. M.; Ho, C.-T. *J. Agric. Food Chem.* **1998**, *46*, 664-667.
27. Hwang, H. I.; Hartman, T. G.; Rosen, R. T.; Lech, J.; Ho, C.-T. *J. Agric. Food Chem.* **1996**, *42*, 1000-1004.
28. Addo, C.; Augustin, J. Changes in the Vitamin B-6 content in potatoes during storage. *J. Food Sci.* **1988**, *53*, 749-752.
29. Saab, R. M. G.; Rao, C. S.; Da Silva, R. S. F. Fortification of bread with L-lysine HCl (hydrochloric acid): losses due to baking process. *J. Food Sci.* **1981**, *46*, 662-663.
30. Selman, J. D. Vitamin retention during blanching of vegetable. *Food Chem.* **1994**, *49*, 137-147.

Chapter 12

Effects of Processing on Tomato Bioactive Volatile Compounds

Mathias K. Sucan[1] and Gerald F. Russell[2]

[1]Applied Food Biotechnology, Inc., 937 Lone Star Drive,
O'Fallon, MO 63366
[2]Department of Food Science and Technology, University
of California, Davis, CA 95616

This study utilized a dynamic headspace sampling with thermal desorption protocol to study the effects of processing on volatile flavor components of tomato juices, and confirmed the presence of numerous previously reported tomato flavor compounds. Aldehydes, alcohols, esters and hydrocarbons were more characteristic of fresh tomato products and decreased in concentration in the heat-treated juice, while the concentrations of sulfur-containing and heterocyclic compounds increased with heat-processing. Results also indicated that volatile substances characterizing fresh tomato juice flavor derive mainly from the metabolism of fatty acids and amino acids, and breakdown of carotenoids. Heat-generated aroma and flavor constituents, on the other hand, arise predominantly from Maillard reactions, caramelization, and from the degradation of lipids, vitamins, pigments and other tomato constituents. The mechanisms of formation of some of these compounds are discussed.

Tomatoes are one of the most consumed food items, both in the fresh and processed states. With recent implications of lycopene in cancer prevention through its ability to act as antioxidant and free radical scavenger (*1,2,3*), tomato consumption is expected to increase. Unlike other carotenoids, lycopene becomes more bioavailable following thermal treatment, and twice and six times as much lycopene were reported in tomato juice and paste, respectively, compared with fresh tomato fruits (5 mg/100g). With consumer increased awareness and demand for healthier foods, consumption of heat-treated tomatoes is bound to rise. Because the characteristic tomato-like flavor is predominantly determined by volatile substances (*4,5,6,7*), studies on the physico-chemical characteristics of flavor compounds in thermally processed tomatoes are still important.

Processed tomato products include whole tomatoes, purees, chili sauce, juice, pulp, paste, catsup and soup (*8*). All these products result from some form of thermal processing during manufacturing. For the manufacture of tomato juice, raw tomatoes are first washed, sorted, and trimmed to remove all visible defects. The tomatoes are then chopped or crushed prior to heating or breaking. The "hot break" method involves heating the product at temperatures exceeding 170°F. Products from this treatment are more viscous and generally regarded as better with respect to cooked tomato flavor and body. In fact, in the hot break procedure, the heat destroys the pectic enzymes and protects the constituents of tomatoes from further enzymatic changes. Lopez (*9*) suggested a hot break condition of 180°F for 15 seconds to destroy all enzymatic activity. Chung and co-workers (*10*) used a hot break condition of 205°F, 22 minutes followed by pasteurization of tomato cans in boiling water for 20 minutes.

After the "hot break" procedure, tomatoes are extracted by pressing against a screen with minimum inclusion of air. A high extraction would yield 3% skins and seeds and 97% juice. Commercially, however, it is common practice to extract only 70% juice, because this low yield contains a higher percentage of soluble solid components which improve flavor, and a lower percentage of insoluble solids which tend to reduce the quality of the finished products.

Tomato juice is sufficiently heat processed either before or after canning to prevent spoilage. In-can sterilization may be carried out at 200° to 205°F for 15 to 20 min. in continuous-agitating cookers. This condition destroys low-heat resistant organisms, but does not provide significant protection against heat resistant spores such as the spores of *Bacillus thermoacidurans* responsible for flat-sour spoilage. Sterilization may be accomplished in bulk before canning. Flash sterilization is the most rapid method employed in canning tomato juice to avoid flat-sour spoilage. Commonly, sterilization is carried out at 250°F for 0.7 min in continuous heat exchangers. The flavor composition of tomato juice is going to be dependent upon the inherent chemical makeup of the fruit, the pH (4.2 for tomato), and the processing conditions. However, since the heat-treatment here is of the type high-temperature short-time process, the conditions are not readily amenable to sufficient non-enzymatic reactions.

Because of the unique popularity of tomatoes in many dishes, tomatoes have been thoroughly studied. Research has already established that tomato flavor is influenced by stages of maturity and ripeness, variety, growth, storage and processing conditions. In general, amino acid-related flavor compounds such as 3-methylbutanal and 2-isobutylthiazole, and glycoside-bound compounds like 6-methyl-5-hepten-2-one, ß-damascenone, furaneol (2,5-dimethyl-4-hydroxy-3(2H)-furanone, and norfuraneol (5-methyl-4-hydroxy-3(2H)-furanone) are formed during ripening. Lipid-derived flavors are generated during maceration while heat-related flavor compounds are produced on cooking. Tomato fruits contain more than 400 volatile compounds (*5,11*), and no single compound can claim responsibility for the characteristic tomato flavor. Heated and unheated tomatoes have distinctive flavor notes.

The effects of processing on bio-active volatile flavor components in tomatoes have been studied with especially noteworthy contributions from Buttery and his co-workers (*12,13*), Sieso and Crouzet (*14*) and Chung et al. (*10*). However, these previous studies did not systematically relate chemical and biochemical mechanisms to quantitative and qualitative determination of tomato flavor compounds, as influenced by processing conditions. Especially the origin of heat-related compounds in fresh tomatoes was not discussed with enough clarity. This was probably be due to the fact that most literature methods used to study tomato volatiles suffer several limitations which may lead to artifact formation, loss of analytes, and proportional changes during sample preparation. For instance, alkyl sulfides and furans are heat-related products by nature. If the analysis conditions used heat during volatile isolation, reporting heat-related compounds in unheated products may be uncomfortable. The goal of this study was to investigate the impact of thermal processing on tomato flavor formation using tomato fruits from the same variety, growth and harvest conditions, and a sensitive, artifact-free and highly reproducible dynamic headspace sampling protocol with thermal desorption.

Methodology

Tomato juices were processed and supplied by Campbell Soup Company R&D center in Davis, CA. Tomato fruits were picked vine-ripe. Part of the fruits was used to make fresh juices, as described by Buttery et al. (*12*). Another portion was heat-processed as tomato juices, using a hot break condition of 180°F, 15 seconds and a sterilization condition of 250°F, 42 seconds. Reference chemicals were reagent grade, supplied from reliable sources.

Preparation of traps

Traps were prepared as described previously (*15,16*). Silane-treated glass tubing (79 mm x 6 mm) was packed with 12 mg 60/80 mesh Tenax-TA (2,6-diphenyl-*p*-phenylene oxide) polymers held in place by silanized glass wool. The traps were initially conditioned at 330°C for 2 hr under a nitrogen gas flow rate of 20 mL/min. Immediately before each purge-and-trap, traps were regenerated at 250°C for 1 hour.

Collection of volatiles

The effects of sample size, purge gas velocity, and purging period on tomato volatile flavor compound composition were previously evaluated (*16*). Triplicate 50 g samples of heat-treated tomato juices or samples of fresh juices with 1 mL internal standard solution (2 ppm 3-pentanone, 3 ppm 2-octanone and 4 ppm anethole), and a magnetic stirring bar were placed in a 500 mL round-bottom flask fitted with a glass sparger. The flask content was brought to 300 g by addition of deionized water. Nitrogen gas (99.997%), at a flow rate of 20 mL/min, was used to purge tomato juice aroma compounds for 60 min. A trap was connected at the gas exit to adsorb the aroma compounds.

Thermal desorption

The recovery of adsorbed volatiles from the trap was done directly inside the GC injector. The required desorption time and temperature were previously evaluated (*15,16*). The injector temperature was 200°C and a loop of the analytical column at the injector end was immersed in a liquid nitrogen-filled Dewar flask to cryogenically trap the desorbed volatiles. Subsequently, the injector glass liner (insert) was replaced with the trap to desorb volatiles. Thermal desorption was carried out for 5 min with the split vent and septum purge closed.

Capillary GC analysis

The oven was initially held at 35°C for 6 min, then linearly increased 5°C/min to 200°C. The GC was equipped with a DB-5 capillary column (50m x 0.32 mm, id) and a flame ionization detector. The detector temperature was 300°C, and the hydrogen carrier gas linear flow rate was 30 cm/sec.

A computer interface was developed between an HP 3392A integrator (Hewlett Packard, Kennett Square, PA) and a Macintosh II computer for capturing and storing chromatographic trace data and integrator report. Software (LabVIEW™, National Instruments) for system automation was developed in this laboratory, and controlled the entire assay system which showed a real-time display of the detector and integrator response (*17*). Chromatographic signals were captured at 20 Hz and stored in ASCII text format to facilitate subsequent analysis by appropriate graphics and analysis applications. Igor™ (Wavemetrics, Beaverton, OR) was used for subsequent chromatographic data treatment and graphing, along with Excel™ (Microsoft) for analyses of integrated peak areas.

Capillary GC-MS analysis

Mass spectra of the GC components from tomato volatiles were obtained on a GC (HP-5890) interfaced to a VG Trio-2 Mass Spectrometer (VG Masslab, Altrincham, UK). The same columns and procedures were used for both GC quantitative and GC-MS qualitative analysis. The carrier gas, however, was helium for GC-MS analyses. Mass spectra were library searched against the

National Institute of Standards and Technology mass spectral reference collection and the Wiley/NBS Registry of Mass Spectra (*18*).

Results and Discussion

The analysis of tomato juice volatile flavors by dynamic headspace sampling method with thermal desorption led to the isolation and quantitation of approximately 150 volatile compounds. This study confirmed the presence of previously identified compounds from tomatoes and tomato products (*6,10,12,19,20,21,22*). Representative chromatograms of volatiles in fresh and heat-processed tomato juices are shown in Figures 1 and 2, respectively. Qualitatively, many identified compounds were identical in both fresh and heated tomato juices. Exceptions include compounds characteristic of either tomato product. On a quantitative basis, however, differences were more noticeable. Aldehyde concentrations decreased considerably and the amounts of alcohols, esters and hydrocarbons decreased moderately during thermal processing. Sulfur-containing and heterocyclic compounds, more characteristic of heated products, and some ketones increased with heat-processing (Table I).

Aldehydes

Aldehydes are known to impart a "green" note to tomato flavor. Results from this study showed that lipid-derived aldehydes, hexanal, (*Z*)-3-hexenal, (*E*)-2-hexenal, (*E*)-2-pentenal, pentanal, heptanal, (*E*)-2-heptenal, octanal and (*E*)-2-octenal were the most abundant aldehydes in fresh tomato juice. The concentrations of these compounds were much lower in the heat-processed tomato juice, due to loss by volatilization during processing. This is in agreement with a previous study by Sieso and Crouzet (*14*) who noted the presence of 3-methylbutanol, (*E*)-2-hexenal, (*Z*)-3-hexenol, (*E*)-2-hexenol and 2-isobutylthiazole in the condensed water from concentrator units used in the manufacture of tomato products. In the present study, the ratio of (*Z*)-3-hexenal to (*E*)-2-hexenal decreased from 7.2 to 0.68, — approximately a 10-fold change in the processed juice. (*Z*)-3-Hexenal was converted to the thermodynamically more stable (*E*)-2-hexenal through isomerization reaction under the heating and acidic pH conditions of the fruit juices. This is consistent with previous work by Chung et al. (*10*) and Hayase et al. (*23*) who did not find (*Z*)-3-hexenal among the volatiles of their tomato juices, and attributed that to isomerization reactions during steam distillation. Karmas et al. (*21*) also found only (*E*)-2-hexenal in fresh tomatoes and thought the isomerization was related to GC conditions. These results indicate that lipid-derived aldehydes were primarily formed in fresh tomato juice and this is consistent with previous findings (*5,24*). In fact, physical disruption of tomato fruit tissues upon maceration results in the release of hydrolytic and oxidative enzymes that degrade the endogenous lipids. Free fatty acids are formed from phospholipids and triglycerides due to the

160

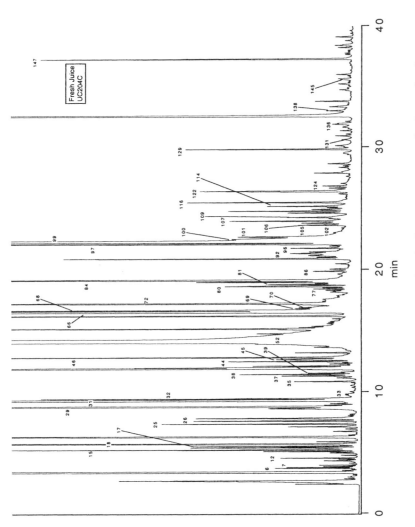

*Figure 1. Chromatogram of volatile flavor compounds from fresh
tomato juice.*

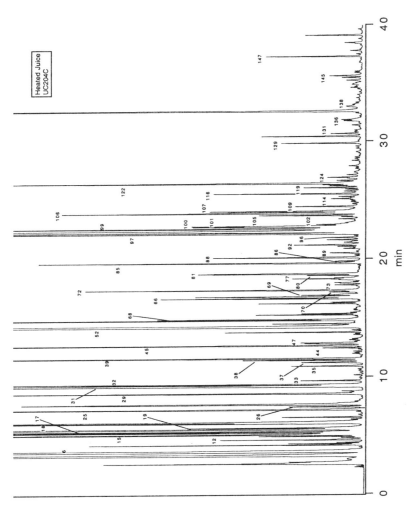

Figure 2. Chromatogram of volatile flavor compounds from heated tomato juice.

Table I. Qualitative and Quantitative Comparison of Fresh and Heated Tomato Juice Volatile Flavors From a Standard Cultivar of *Lycopersicon esculentum*.

FRESH TOMATO JUICE VOLATILES		Peak#	HEATED TOMATO JUICE VOLATILES	
Conc. (ppm)	Compound		Compound	Conc. (ppm)
	ALDEHYDES		ALDEHYDES	
0.20	2-Methyl-2-butenal	17	2-Methyl-2-butenal	0.62
0.24	3-Methylbutanal	25	3-Methylbutanal	0.75
2.29	(E)-2-pentenal	29	(E)-2-pentenal	0.28
5.09	Pentanal	31	Pentanal	0.47
83.91	Hexanal	52	Hexanal	6.21
11.88	(Z)-3-hexenal	66	(Z)-3-hexenal	0.23
1.64	(E)-2-hexenal	72	(E)-2-hexenal	0.34
1.68	Heptanal	88	(E)-2-heptenal	0.18
0.53	(E)-2-heptenal	89	Benzaldehyde	0.03
0.40	Octanal	101	Octanal	0.23
0.08	Phenylacetaldehyde	105	Phenylacetaldehyde	0.12
0.27	(E)-2-octenal	116	(E)-2-octenal	0.14
0.40	Decanal	122	Decanal	0.55
0.07	Undecanal	131	Undecanal	0.19
0.13	Geranial	136	Geranial	0.13
0.09	2,4-Decadienal	138	2,4-Decadienal	0.16

KETONES

2-Butanone	0.09
1-Penten-3-one	0.08
6-Methyl-5-hepten-2-one	1.29
ß-Damascenone	0.11
Geranylacetone	0.80

ALCOHOLS

3-Methylbutanol	0.10
Pentanol	0.16
(Z)-3-hexenol	0.75
1-Hepten-3-ol	0.08
6-Methyl-5-hepten-2-ol	0.23
2-Phenylethanol	0.29
2-Propylphenol	0.12

KETONES

2-Butanone	12	0.13
Acetoin	19	0.22
1-Penten-3-one	35	0.10
6-Methyl-5-hepten-2-one	97	1.33
ß-Damascenone	145	0.22
Geranylacetone	147	0.50

ALCOHOLS

3-Methylbutanol	37	0.09
Pentanol	44	0.04
(Z)-3-hexenol	68	0.26
2,4-Dimethylphenol	85	0.51
1-Hepten-3-ol	96	0.04
6-Methyl-5-hepten-2-ol	100	0.23
2-Phenylethanol	107	0.25
2-Propylphenol	109	0.08
a-Terpineol	119	0.06

Continued on next page.

Table I. Continued.

FRESH TOMATO JUICE VOLATILES			HEATED TOMATO JUICE VOLATILES	
Conc. (ppm)	Compound	Peak#	Compound	Conc. (ppm)
	ESTERS		ESTERS	
0.05	Methyl acetate	7		
0.25	Methyl isobutyrate	46		
0.06	Methyl hexanoate	86	Methyl hexanoate	0.03
		102	Hexyl acetate	0.09
0.04	Methyl octanoate	124	Methyl octanoate	0.04
0.29	Methyl salicylate	129	Methyl salicylate	0.07
	SULFUR COMPOUNDS		SULFUR COMPOUNDS	
0.08	Dimethyl sulfide	6	Dimethyl sulfide	29.74
0.05	Dimethyl disulfide	39	Dimethyl disulfide	0.63
0.10	Dimethyl trisulfide	92	Dimethyl trisulfide	0.09

	HETEROCYCLICS	
0.39	2-Methylfuran	
1.82	3-Methylfuran	
0.51	2-Ethylfuran	
trace	2,5-Dimethylfuran	
0.06	Ethylthiophene	
3.64	2-Pentylfuran	
0.64	2-Isobutylthiazole	
	HYDROCARBONS	
0.17	Benzene	
0.14	2-Pentene	
0.26	Toluene	
0.05	Xylene	
0.03	2,4-Dimethylhexene	
0.30	Styrene	
0.12	Nonane	
0.25	4-Methyl-1,5-heptadiene	

	HETEROCYCLICS	
15	2-Methylfuran	1.06
18	3-Methylfuran	3.20
32	2-Ethylfuran	0.49
33	2,5-Dimethylfuran	0.08
47	Methylthiophene	0.08
69	Ethylthiophene	0.08
73	2-Acetylfuran	0.03
99	2-Pentylfuran	0.70
106	2-Isobutylthiazole	0.45
	HYDROCARBONS	
26	Benzene	0.10
38	2-Pentene	0.15
45	Toluene	0.61
70	Xylene	0.03
77	2,4-Dimethylhexene	0.06
80	Styrene	0.05
81	Nonane	0.27
114	4-Methyl-1,5-heptadiene	0.03

actions of acyl hydrolase and phospholipase D (5). Mainly the free fatty acid linoleates and linolenates are subsequently oxidized by lipoxygenases, and the resulting hydroperoxides are specifically cleaved by hydroperoxidases to form C6 alcohols and aldehydes. Lipid-related 2,4-Decadienal was also present in fresh juice, and confirmed previous report by Buttery and Ling (24) and Karmas et al. (21). Moreover, 2,4-decadienal stems from the degradation of the 9-hydroperoxide isomer, in addition to literature report that the 13-hydroperoxide isomer is cleaved by tomato enzymes into C6 alcohols and aldehydes.

Aldehydes can also originate from amino acid metabolism in fresh juice. A crude enzyme solution prepared from green tomatoes formed 3-methylbutanal and 3-methylbutanol when leucine was used as a substrate (25, 26). It is possible that aldehydes are formed from amino acids by transamination reactions, a process similar to that in banana. The initial step in the reaction series is deamination of the amino acid followed by decarboxylation. A series of reduction and esterification reactions then follow and lead to a number of volatiles significant to flavor. In the present study, the amino acid-related aldehydes 3-methylbutanal and 2-methyl-2-butenal did not appear to be affected by the processing conditions. Rather, their concentrations increased in the heated juice (Table I). A portion of these compounds may have originated from Strecker degradation during thermal processing. Strecker aldehydes include 3-methylbutanal from leucine, methional from methionine and phenylacetaldehyde from phenylalanine. The concentration of phenylacetaldehyde increased about 1.5 % in the heat-processed juice, which is consistent with the 1.2 % increase reported by Buttery and co-workers (20). This compound may be also formed during processing, probably through glycoside hydrolysis.

Geranial, which survived the processing conditions, is a carotenoid-related aldehyde. Most of the carotenoid-related compounds may be formed from the degradation of carotenoids by enzymatic reactions or coupled oxidation reactions of linoleate catalyzed by lipoxygenase (27).

Ketones

Ketones elicit a fruit-like aroma in tomato juices. They are more stable to processing conditions than aldehydes. In the present work, we identified 2-butanone, 1-penten-3-one, 6-methyl-5-hepten-2-one, ß-damascenone and geranylacetone. It is not clear how 2-butanone is formed in fresh tomatoes, but one possible pathway may be via fatty acid metabolism during ripening (28). 1-Penten-3-one was reported among the products of fatty acid oxidation by lipoxygenase (24).

Geranylacetone, ß-damascenone and 6-methyl-5-hepten-2-one have been reported to be carotenoid-related volatile compounds. The amount of 6-methyl-5-hepten-2-one increased in the heat treated tomato juice while the concentration of geranylacetone decreased. This is in agreement with study of Buttery and co-workers (29) who reported that 6-methyl-5-hepten-2-one was found among the products of thermal hydrolysis of tomato glycosides while geranylacetone was absent. Therefore, one pathway of 6-methyl-5-hepten-2-one formation may be through glycoside hydrolysis while the principal mode of geranylacetone formation may be by oxidative processes coupled with lipid oxidation.

The concentration of ß-damascenone doubled during heat-processing. This is consistent with results from studies by Buttery et al. (20). The main mode of ß-damascenone formation has been reported to be through glycoside hydrolysis. In fresh tomato juice, enzyme systems may be responsible for the hydrolysis of ß-damascenone glycosides. In heated juice, however, chemical hydrolysis may predominate (11,29).

Alcohols

With the exception of linalool, the contribution of aliphatic alcohols to tomato aroma is rather low based on their odor unit values as determined by Buttery and co-workers (20). The loss of alcohols through evaporation during thermal processing was more extensive than that of ketones, but less than that of aldehydes. Most of the alcohols may have resulted from reductase conversions of the corresponding aldehydes or ketones formed from the metabolism of fatty acids and amino acids, and from the degradation of carotenoids.

Esters

The flavor characteristics elicited by esters in tomatoes has not been clearly defined, although esters are important contributors to the aroma of fruits such as banana and pears. For example, the decadienoate esters are generally considered carrier of the flavor of pears and elicit fruity and sweety notes (30). Some esters such as ethyl hexanoate, hexyl acetate, ethyl heptanoate, ethyl octanoate and ethyl decanoate were described as sweet, ester and fruity upon sniffing GC eluates of beers (31). The present study of tomato volatile flavors revealed the presence of methyl acetate, hexyl acetate, methyl octanoate and methyl salicylate at low concentrations compared to the other component aldehydes and ketones. A comparison of the amount of esters in fresh and heat-processed tomato juices indicated that there was loss of esters, mainly methyl salicylate, during processing. Methyl salicylate — a lignin-related volatile formed through the shikimic acid pathway — is bound as a glycoside in tomatoes.

The aliphatic esters may be products of fatty acid metabolism (ß-oxidation) during the ripening of fruits. Once ripening commences, this metabolic pathway functions to produce numerous fatty acids, ketones, aldehydes and alcohols, which can be converted to various esters by endogenous enzyme systems (30).

Sulfur-containing compounds

Sulfur-containing compounds are known to contribute to the aroma of cooked foods. However, the acyclic sulfur-containing compounds dimethyl sulfide, dimethyl disulfide and dimethyl trisulfide, identified in this study, were all found in both fresh and heat-treated tomato juices. Other studies have reported dimethyl sulfide (*10*) and dimethyl disulfide (*10, 21, 23*) in fresh tomato juice. Buttery et al (*24*) reported the presence of dimethyl trisulfide in tomato paste, and found that the odor threshold for this compounds was very low (0.01 ppb) compared to that of dimethyl sulfide (0.3 ppb). The occurrence of dimethyl trisulfide in fresh tomato juice was reported by Karmas et al. (*21*). Results also showed that the amount of dimethyl sulfide increased steeply from 0.08 ppm to 29.7 ppm, — about a 370-fold increase. Dimethyl disulfide increased only about 12-fold and dimethyl trisulfide concentration remained constant during thermal processing.

The major mode of dimethyl sulfide formation is through heat-treatment. As an effect of heat, the S-methylmethionine sulfonium ion breaks down into dimethyl sulfide and homoserine (*6,20,32*). Part of dimethyl disulfide formation may have also occurred during heat-treatment, probably from the thermal decomposition of sulfur-containing amino acid precursors. The amount of dimethyl trisulfide did not change appreciably during the manufacture of tomato juices. This indicates that the process of dimethyl trisulfide formation may take place almost entirely in fresh juice.

The formation of dimethyl sulfide, dimethyl disulfide and dimethyl trisulfide in fresh tomatoes is probably through pathways other than the thermal breakdown of sulfur-containing precursors. The biogenesis of these sulfides in fresh tomatoes has not yet been elucidated. However, it is possible that the formation of sulfides in fresh tomato juice derived from the breakdown of methionine or S-methylmethionine sulfonium ion. UV radiation exposure of methionine and possibly its methyl sulfonium ion derivative, in the presence of riboflavin and oxygen, generates methional, which further degrades to methanethiol, dimethyl sulfide, dimethyl disulfide (*33*), and dimethyl trisulfide.

Heterocyclics

The heterocyclic compounds containing oxygen nitrogen, sulfur or combinations of these atoms generally result from Maillard reactions during thermal processing, and are well known for their contribution to the aroma of cooked foods. The process of Maillard reaction at pHs lower than 5 (as in the case of tomatoes) promotes the formation of furan derivatives, pyrones and thiophenes. In the present study, we were able to identify several furan derivatives, two thiophenes and a thiazole. Chung and co-workers (*10*) reported that pyrroles and pyrazines were not detected in tomato juices, but in purees and paste.

Ethylthiophene was identified in fresh tomato juice in this study. Although 2-formylthiophene was characterized in fresh tomato juice by Ho and Ichimura (34), the pathway of formation of thiophene derivatives in unheated tomatoes has not yet been proposed. Thiophenes occur naturally as secondary metabolites in Marigold plants, and are known for their phototoxicity (35).

Furans are also characteristic of heated products. However, our results showed the presence, in fresh juice, of furans in low concentrations, which increased during processing. Other studies have reported the presence of 2-methylfuran, 2-ethylfuran, 2-pentylfuran and 2-acetylfuran in fresh tomato juices (21,23). The origin of these furans in fresh juice may be through lipid oxidation reactions. Lipid oxidation can involve the reaction of molecular oxygen with unsaturated fatty acids via a free radical mechanism. This self-propagating reaction leads to the formation of unstable hydroperoxides, which undergo scission at the oxygen-oxygen bond to generate alkoxy radicals. These alkoxy radicals may serve as precursors for the formation of furans. In this study, increasing amount of furans during heat-processing was probably due to both lipid oxidation and nonenzymatic browning.

2-Isobutylthiazole is the only alkylthiazole often found in tomatoes, and was previously reported to be formed during the ripening of the tomato fruit; see for example Petro-Turza (5), Buttery et al. (12,19,20) and Baldwin et al. (6). Hayase et al. (23) found this compound to have a grassy and sweet fruity odor in sensory evaluation by GC-sniffing analyses. Results from the present study showed that the amount of 2-isobutylthiazole decreased during heat processing, which is consistent with the report by Buttery et al. (20) and Sieso and Crouzet (14) who found a considerable loss of 2-isobutylthiazole during tomato processing. The pathway of 2-isobutylthiazole formation has not yet been clearly elucidated. However, one hypothesis suggests the formation of 3-methylbutanal which subsequently condenses with cysteamine to give a thiazolidine. The thiazolidine then can oxidize to thiazole (24,36).

2-Isopropyl-4-methylthiazole, 4-butyloxazole and 4-ethyl-5pentyloxazole were reported for the first time in fresh tomatoes by Ho and Ichimura (34).

Hydrocarbons

The amounts of the hydrocarbons benzene, xylene and styrene decreased during heat-treatment, which is an indication that these compounds may be exclusively generated in the fresh juice, and may not be strongly characteristic of heated tomato juices. Toluene, however, increased with thermal processing, and this is in agreement with the work by Sieso and Crouzet (14). The importance of these compounds to tomato flavor still remain to be determined.

Conclusions

This study confirmed the presence of previously identified compounds in tomatoes. The concentrations of aldehydes, alcohols, esters and hydrocarbons, more characteristic of fresh tomatoes, decreased in the heat-processed juices probably due to loss by evaporation and/or degradation reactions. (Z)-3-hexenal was found in extremely low concentration (0.23 ppm) in the heated juice probably due to thermodynamically-driven isomerization to (E)-2-hexenal. The occurence of lipid-related 2,4-Decadienal in fresh juice may suggest that the 9-hydroperoxide isomer was also cleaved during the oxidation of linoleates and linolenates, in addition to the cleavage of 13-hydroperoxide isomer. Compounds more characteristic to heated products such as sulfur-containing and heterocyclic compounds increased in concentrations with processing. The occurrence of dimethyl sulfide, dimethyl disulfide and dimethyl trisulfide in fresh juice may be through other pathways than the thermal breakdown of sulfur-containing precursors. Although the mechanisms of volatile flavor formation was rationalized in this study, more work is needed to elucidate the biogenesis of the sulfides, the furan derivatives and the esters in fresh tomatoes.

Acknowledgments

The authors are indebted to Dr. Dan Jones of Facility for Advanced Instrumentation (FAI), University of California Davis, for his help with the operation of the quadrupole mass spectrometer, and for his availability for questions related to the interpretation of mass spectral data. We wish to acknowledge, with great appreciation, the technical support from the research staff at J&W Scientific who offered many helpful suggestions during the course of this research and the donation (grant-in-aid) of a high efficiency fused silica column by J&W Scientific in support of graduate student research.

We also thank Dr. Joseph DeVerna and Dawn Adams of the Campbell Research and Development Center in Davis, CA for supplying us with standard cultivars, wild species and hybrids of tomatoes, during the course of this investigation.

References

1. Giovannuci, E.; Ascherio, A.; Rimm, E. B.; Stampfer, M. J.; Colditz, G. A.; Willett, W. C. *J. National Cancer Institute* **1995**; *87*, 1767-76.
2. Khachik, F.; Beecher, G. R.; Smith, J. C. Jr. *J. Cell Biochem.* Suppl. **1995**, *22*, 236-246.
3. Nguyen, M. L.; Schwartz, S. *J. Food Technology* **1999**, *53*, 38-45.
4. Kopeliovitch, E.; Mizrahi, Y.; Rabinowitch, H. D.; Kedar, N. *J. Amer.Soc. Hort. Sci.* **1982**, *107*, 361.
5. Petro-Turza, M. *Food Reviews International* **1987**, *2*, 309.

6.   Baldwin, E. A.; Nisperos-Carriedo, M. O.; Baker, R.; Scott, J. W. *J of Agric. and Food Chem.* **1991**, *39*, 1135.
7.   Baldwin, E. A; Scott, J. W.; Einstein, M. A.; Malundo, T. M.; Carr, B. T.; Shewfelt, R. L.; Tandon, K. S. *J. Amer. Soc. Hort. Sci.* **1998**, *123*, 906-915.
8.   Gould, W. A. Tomato production, processing and technology; CTI Publications, Baltimore, MD, 1992.
9.   Lopez, A. A Complete Course in Canning. Canning Trade, Baltimore, MD, 1987.
10.   Chung, T. Y.; Hayase, F.; Kato, H. *Agricultural and Biological Chemistry* **1983**, *47*, 343.
11.   Marlatt, C.; Chien, M.; Ho, C. T. *Journal of Essential Oil Research*, **1991**, *3*, 27.
12.   Buttery, R. G.; Teranishi, R.; Ling, L. C. *J. of Agric. and Food Chem.* **1987**, *35*, 540.
13.   Buttery, R. G.; Takeoka, G. R.; Krammer, G. E.; Ling, L. C. *Lebensm. Wiss. Technol.* **1994**, *27*, 592-594.
14.   Sieso, V.; Crouzet, J. *Food Chemistry* **1977**, *2*, 241.
15.   Sucan, M. K. Ph.D., Dissertation, University of California, Davis, CA, 1995.
16.   Sucan, M. K.; Russell, G. F. *J. High Resol. Chromatogr.* **1997**, *20*, 310.
17.   Budwig, C. E. *Personal Communication* 1995.
18.   McLafferty, F. W.; Stauffer, D. B. The Wiley/NBS Registry of Mass Spectral Data; John Wiley & Sons: New York, 1989.
19.   Buttery, R. G.; Teranishi, R.; Ling, L. C and Stern, D. J. *J. Agr Food Chem.* **1988**, *36*, 1247.
20.   Buttery, R. G.; Teranishi, R.; Ling, L. C.; Turnbaugh, J. G. *J. Agric. Food Chem.* **1990**, *38*, 336.
21.   Karmas, K.; Hartman, T. G.; Salinas, J. P.; Lech, J.; Rosen, R. T. In *Lipids in Food Flavors*; Ho, C.-T.; Hartman, T. G., Eds.; ACS Symposium Series: Washington, DC, 1994; pp 130.
22.   Krumbein, A.; Ulrich, D. In *Flavour Science*: Recent Revelopments; A. J. Taylor, D. S. Mottram, Eds.; The Royal Society of Chemistry: Cambridge, UK, 1996; pp 289.
23.   Hayase, F.; Chung, T.; Kato, H. *Food Chemistry* **1984**, *14*, 113.
24.   Buttery, R. G.; Ling, L. C. American Chemical Society: Washington, D. C., 1993; Vol. ACS Symposium Series 525.
25.   Yu, M.-H.; Olson, L. E.; Salunkhe, D. K. *Phytochemistry* **1968**, *7*, 555
26.   Yu, M.-H.; Olson, L. E.; Salunkhe, D. K. *Phytochemistry* **1968**, *7*, 561.
27.   Stevens, M. A. *J. Amer. Soc. Hort. Sci.* **1970**, *95*, 461.
28.   Schreier, P. Chromatographic Studies of Biogenesis of Plant Volatiles. W. Bertsch, W. G. Jennings, R. E. Kaiser, Eds.; Verlag: New York, 1984.

29. Buttery, R. G.; Takeoka, G.; Teranishi, R.; Ling, L. C. *J. Agric. Food Chem.* **1990**, *38*, 2050.
30. Heath, H. B.; Reineccius, G. *Flavor Chemistry and Technology*; AVI Publishing Company, Inc. Westport: Connecticut, 1986.
31. Fors, B. S. M.; Nordlon, H. *J. Institute of Brewing* **1987**, *93*, 496.
32. Wong, F. F.; Carson, J. F. *J. Agric. Food Chem.* **1966**, *14*, 247 .
33. Marsilli, R. In *Techniques For Analyzing Food Aroma*. R. Marsili, ed.; Marcel Dekker, Inc., New York, NY,1997.
34. Ho, C. T.; Ichimura, N. *Lebensmittel Wissenschaft und Technologie* **1982**, *15*, 340.
35. Wells, C. *Chromatographia* **1992**, *34*, 241-248.
36. Stone, E. J.; Hall, R. M.; Kazeniac, S. J. Abstracts of Papers. *American Chemical Society* **1971**, *162*, 31.

Chapter 13

Formation of Bioactive Peptides from Milk Proteins through Fermentation by Dairy Starters

Hannu Korhonen and Anne Pihlanto-Leppälä

Agricultural Research Centre of Finland, Food Research,
FIN–31600 Jokioinen, Finland

Milk proteins have been identified as an important source of bioactive peptides. Such peptides can be released during hydrolysis induced by digestive or microbial enzymes. The peptides derived from milk proteins have been shown to have a variety of different functions *in vitro* and *in vivo,* e.g., opioid, antihypertensive, antimicrobial, antithrombotic, immunomodulatory or mineral-carrying. There is commercial interest in the production of bioactive peptides with the purpose of using them as active ingredients in functional foods. Industrial scale production of such peptides is, however, hampered by the lack of suitable technologies. Bioactive peptides can also be produced from milk proteins through the fermentation of milk by starters employed in the manufacture of fermented milks or cheese. The formation during milk fermentation of a number of peptides with various bioactivities has been observed in many recent studies. Antihypertensive peptides have been identified in fermented milk, whey and ripened cheese. A few of these peptides have proved effective also in humans after ingestion of fermented milk. In our own studies, caseins hydrolyzed with a probiotic *Lactobacillus* strain and digestive enzymes generated a number of immunomodulatory peptides. Fermentation of milk with the same strain followed by digestion with gastric enzymes resulted in a spectrum of bioactive peptides, e.g., immunomodulatory, opioid and antihypertensive. Such peptides may contribute to the reputed health-promoting properties of fermented dairy products.

The release of biologically active peptides from milk proteins was first reported more than fifty years ago. Since then, milk proteins have been identified as an important source of bioactive peptides (Table I). These peptides are in an inactive state inside the protein molecule and can be released during enzymatic digestion *in vitro* or *in vivo*. Bioactive peptides usually contain 3 to 20 amino acid residues per molecule. Biologically active peptides have been found to have specific activities, such as antihypertensive, antimicrobial, immunomodulatory, opioid or mineral-binding. Many milk-derived peptides reveal multifunctional properties, i.e., specific peptide sequences may exert two or more different biological activities. In recent years, many review articles have been published about bioactive peptides, which have been regarded as highly prominent ingredients for application in health-promoting functional foods or pharmaceutical preparations (*1-4*).

Peptides with biological activity can be produced in three ways: a) enzymatic hydrolysis with digestive enzymes, b) fermentation of a food material with proteolytic microorganisms or c) through the action of enzymes derived from proteolytic microorganisms. Natural or controlled fermentation involves microorganisms (starter cultures), e.g., lactic acid bacteria, which degrade proteins during their growth in their surrounding medium, resulting in a variety of peptides with various amino acid sequences and single amino acids. The degree of proteolysis is dependent on the bacterial species involved as well as the physical conditions of fermentation (*5*).

Over the last ten years, an increasing number of studies indicate that during fermentation of milk with certain dairy starters, peptides with various bioactivities are formed. These can be traced in an active form even in the final products, such as fermented milks and cheese (*6*).

This article will review the current knowledge about bioactive peptides derived from milk proteins with a special emphasis on formation of such peptides through microbial fermentation. Recent results from our own studies are also reviewed.

Table I. Milestones in the Discovery of Milk-derived Bioactive Peptides

Decade of discovery	Bioactive peptides	Protein precursor	Bioactivity	Ref
1950	Phosphopeptides	α- and β-Casein	Mineral carrier	45
1970	Casomorphins	β-Casein	Opioid agonist	46
	α-Casein exorphin	α-Casein	Opioid agonist	47
1980	Immunopeptides	α- and β-Casein	Immunostimulant	48
	Casokinins	α- and β-Casein	ACE-inhibitory	49
	Casoxins	κ-Casein	Opioid antagonist	50
	Lactorphins	α-Lactalbumin and β-Lactoglobulin	Opioid agonist	51
	Casoplatelins	κ-Casein	Antithrombotic	52
1990	Lactoferricins	Lactoferrin	Antimicrobial	53
	Lactokinins	α-Lactalbumin and β-Lactoglobulin	ACE-inhibitory	54

Formation of Bioactive Peptides by Microbial Fermentation

Many of the traditionally used dairy starter cultures are highly proteolytic, and the peptides and amino acids degraded from milk proteins contribute to the typical flavor, aroma and texture of the products.

The proteolytic system of lactic acid bacteria, such as *Lactococcus lactis*, *Lactobacillus helveticus* and *Lactobacillus delbrueckii* var. *bulgaricus*, is already well known (*7-10*). This system consists of a cell wall bound proteinase and several intracellular peptidases. In recent years, rapid progress has been made in the elucidation of the biochemical and genetic characterization of these enzymes (*11-13*). The proteolytic system of lactic acid bacteria provides transport systems specific for amino acids, di- and tripeptides and oligopeptides up to 18 amino acids. Longer oligopeptides which are not transported into the cells can be a source for the liberation of bioactive peptides in fermented milk products when further degraded by intracellular peptidases after cell lysis (*9*). In the gastrointestinal tract, digestive enzymes may further degrade long oligopeptides, leading to a possible release of bioactive peptides. Once liberated in the intestine, bioactive peptides may act locally or pass through the intestinal wall into blood circulation ending up in a target organ, with subsequent regulation of physiological conditions through, e.g., nerve, immune, vascular or endocrine systems. Figure 1 illustrates potential *in vivo* functions of different bioactive peptides derived from fermented milk products.

Hamel et al. *(14)* observed β-casomorphin immunoreactive material in cow's milk incubated with caseolytic bacteria, e.g., *Pseudomonas aeruginosa* or *Bacillus cereus*. Matar and Goulet *(15)* detected β-casomorphin-4 in milk fermented with *L. helveticus* L89. Pihlanto-Leppälä et al. *(16)* studied the potential formation of angiotensin converting enzyme (ACE)-inhibitory peptides from cheese whey and caseins in fermentation with different commercial lactic acid starters used in the manufacture of yoghurt, ropy milk and soured milk. The proteolytic activity of all starters was weak and the rate of proteolysis was only 1–3% during fermentation. No ACE-inhibitory activity was observed in these hydrolysates. Further digestion of above samples with pepsin and trypsin increased the proteolysis rate to about 45% for whey proteins and 70% for caseins. This resulted in an ACE-inhibition rate varying between 35–61% for the whey protein hydrolysates and 86% for the casein hydrolysates, respectively. Among whey hydrolysates, peptides with small molecular weight (< 1000 Da) were the most active inhibitors, and several active peptides derived from α-lactalbumin and β-lactoglobulin were identified. Among casein hydrolysates, ACE-inhibitory peptides were released both from $α_{s1}$- and β-caseins.

Belem et al. *(17)* fermented whey with *Kluyveromyces marxianus* var. *marxianus,* and identified in the hydrolysate a tetra peptide which had a sequence of β-lactorphin (Tyr-Leu-Leu-Phe). It was suggested that this peptide may have antihypertensive properties. However, no *in vivo* studies with this peptide have been so far reported.

Formation of Bioactive Peptides by Microbial Enzymes

Microbial enzymes have been used successfully to produce bioactive peptides from milk proteins. Yamamoto et al. *(18)* reported that casein hydrolyzed by the cell wall-associated proteinase from *L. helveticus* CP790 showed antihypertensive activity in spontaneously hypertensive rats (SHR). Several ACE-inhibitory and one antihypertensive peptide were isolated from the hydrolysate. Using the same proteinase, Maeno et al. *(19)* identified a β-

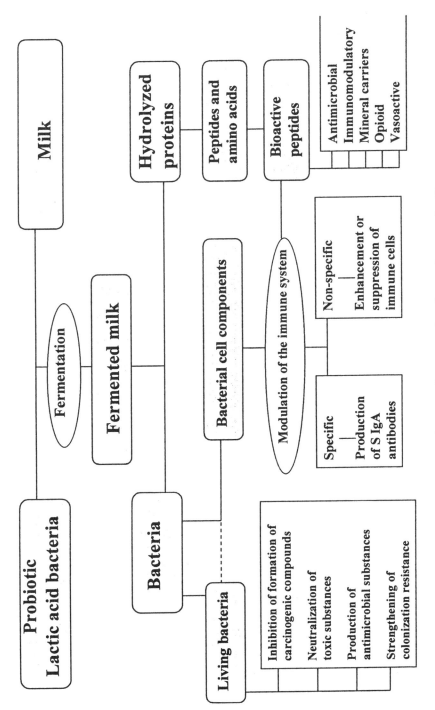

Figure 1. Potential in vivo functions of probiotic fermented milks.

casein-derived antihypertensive peptide from the casein hydrolysate. The antihypertensive effect of this peptide (Lys-Val-Leu-Pro-Val-Pro-Glu) was dose-dependent in SHR at a dosage level from 0.2 to 2 mg of peptide per kg body weight. This peptide did not show a strong ACE-inhibitory activity as such but a corresponding synthetic hexa-peptide, deleted by Glu (Lys-Val-Leu-Pro-Val-Pro) had a strong ACE-inhibitory activity, as well as a significant antihypertensive effect in SHR. It was suggested by the researchers that both the proline residue in the C-terminus and the amino acid sequence might be important for ACE-inhibition of the hexapeptide. Proline is generally known to be resistant to degradation by digestive enzymes.

Abubakar et al. (20) observed that cheese whey digested with proteinase K from *Tritirachium album* had a depressive effect on the systolic blood pressure of SHR, with the highest antihypertensive activity found with the tripeptide Ile-Pro-Ala as derived from β-lactoglobulin.

A few *in vitro* studies have shown that bioactive peptides derived from milk proteins after hydrolysis with digestive and/or microbial enzymes modulate the proliferative activity of human peripheral blood lymphocytes. Sütas et al. (21) demonstrated that digestion of casein fractions by pepsin and trypsin produced peptides which had either immunostimulatory or immunosuppressive influence on human blood lymphocytes. Peptides derived from total casein and α_{s1}-casein were mainly suppressive, while those derived from β- and κ-casein were primarily stimulatory. When the caseins were hydrolyzed by enzymes isolated from a probiotic *Lactobacillus GG* var. *casei* strain prior to pepsin-trypsin treatment, all hydrolysate fractions obtained from HPLC runs were immunosuppressive, and the highest activity was again found in α_{s1}-casein (Table II). These results suggest that lactic acid bacteria may modulate the immunogenicity of milk proteins prior or after oral ingestion of the product. Such a modulation by fermentative bacteria may be beneficial in the down regulation of hypersensitivity reactions to ingested proteins in patients with food protein allergy. Further studies (22) showed that the above casein hydrolysates modulated the *in vitro* production of cytokines by human blood T lymphocytes. Cytokines are associated with the regulation of allergic reactions.

Indeed, promising results have been obtained in the management of atopic reactions of infants by oral bacteriotherapy with the probiotic *Lactobacillus GG* strain (23). These results are also supported in a study by Laffieneur et al. (24), who showed that β-casein hydrolyzed by lactic acid bacteria, e.g., *L. helveticus*, has immunomodulatory activity which could be related to interaction with monocyte-macrophage and T-helper cells.

Table II. Immunogenicity of Casein-hydrolysates after Treatment of Casein Components with Pepsin-trypsin or Lactobacillus GG Enzymes Followed by Pepsin-trypsin

Casein component	Immunomodulatory effect [a]	
	Pepsin-trypsin	L. GG + pepsin-trypsin
Total casein	suppressive	suppressive
α_{s1}-casein	suppressive	suppressive
β-casein	stimulatory	suppressive
κ-casein	stimulatory	suppressive

(Data compiled from Sütas et al. (22))

[a] human blood lymphocyte proliferation test applied

Occurrence of Bioactive Peptides in Fermented Milk Products

Many studies have reported on the formation or presence of various bioactive peptides in fermented milk products. Such peptides have been identified in sour milk upon fermentation with strong proteolytic starter cultures, e.g., *L. helveticus*, and in ripened cheese varieties (Table III). The formation of calcium-binding phosphopeptides in fermented milks and various types of cheese has been observed in many studies (25-27). The physiological role of these peptides is still controversial. β-casomorphins have been detected in fermented milk (28,15), but in ripened cheese only precursors of β-casomorphins have been identified (29,30).

Rokka et al. (31) reported the release of a variety of bioactive peptides by enzymatic proteolysis of UHT milk fermented with a probiotic *Lactobacillus casei* ssp. *rhamnosus* strain. Upon fermentation, the product was treated with pepsin and trypsin with the intent to simulate gastrointestinal conditions. In the hydrolysate, many bioactive peptides, e.g., ACE-inhibitory, immunomodulatory and opioid peptides, were identified, this suggesting that these bioactive peptides may partially explain the health-conducive properties of probiotic bacteria. A similar suggestion was made by Moineau and Goulet (32), who studied the effect of feeding *Lactobacillus helveticus* fermented milk on the pulmonary macrophage activity in mice.

Meisel et al. (33) analyzed a variety of dairy products, sports nutrients and infant formulas for the presence of ACE-inhibitory peptides. Low ACE-

Table III. Bioactive Peptides Identified in Fermented Milk Products

Product	Bioactivity observed	Ref
Sour milk	Phosphopeptides	25
Sour milk	Antihypertensive	41,42
Sour milk	β-casomorphin-4	15,28
Fermented milk (treated with pepsin and trypsin)	ACE-inhibitory Immuomodulatory Opioid	31
Yoghurt	ACE-inhibitory (weak)	33
Quarg	ACE-inhibitory	33
Cheese (Parmesan Reggiano)	β-casomorphin precursors	29
Cheese	No β-casomorphin	30
Cheese (Comté)	Phosphopeptides (CPP)	26
Cheese (Cheddar)	Phosphopeptides (CPP)	27
Cheese (e.g. Edam, Emmental, Gouda, Roquefort, Tilsit)	ACE-inhibitory	33
Cheese (Italian varieties)	ACE-inhibitory	55
Cheese (Finnish varieties: Edam, Emmental, Turunmaa, Cheddar, Festivo)	ACE-inhibitory	34

inhibitory activity was measurable in samples having a low degree of proteolysis, e.g., yoghurt, fresh cheese, quarg and sports nutrition-related products. In ripened cheese types, the inhibitory activity increased with developing proteolysis but started decreasing when cheese maturation exceeded a certain level, as measured by the free peptide-bound amino acids ratio.

The above results were supported by our analyzes in regard to the occurrence of ACE-inhibitory peptides in Finnish cheese varieties (unpublished results). These peptides were found in all tested samples of fermented cheese, but the level varied from one cheese type to another as well as in relation to the age of cheese. These results would suggest that ACE-inhibitory and probably also other biologically active peptides are naturally formed in cheese, remain active for a limited period and then are split into other inactive peptides and amino acids as the ripening proceeds. This hypothesis was confirmed in our further studies, in which we followed the formation of ACE-inhibitory peptides in a novel probiotic cheese type that we had developed (34). This Gouda-like hard cheese was fermented with *Lactobacillus acidophilus* and *Bifidobacteria* in addition to cheese starters. ACE-inhibitory activity was first noted in 6-

week-old cheese, reaching the highest level at the age of 13 weeks. Thereafter, the activity began decreasing slowly. The role of specific starter cultures in this process remains to be elucidated, as, e.g., ACE-inhibitory peptides seem to be regularly formed in different cheese varieties manufactured using various starter organisms. Also, the importance in cheese of non-starter lactic acid bacteria for the formation and transformation of bioactive peptides remains to be studied.

Biological Role of Bioactive Peptides

The potential physiological role of bioactive peptides occurring naturally in fermented dairy products and the *in situ* formation of these peptides in the gastrointestinal tract is unknown. There are already a number of *in vivo* studies (Table IV) which demonstrate that peptides with various activities can be found in an active form in blood circulation as well as in certain target organs upon oral administration of pure peptides or their precursors (*35-40*).

In animal and human studies, the physiological role of antihypertensive ACE-inhibitory peptides is most studied, so far. Substantial experimental evidence is available about the antihypertensive properties of a Japanese sour milk called Calpis. This product has been manufactured commercially in Japan for more than 80 years. In the preparation of Calpis, a starter culture containing *Lactobacillus helveticus* and *Saccharomyces cerevisieae* is used. During fermentation, two specific tripeptides namely Val-Pro-Pro and Ile-Pro-Pro are formed from β-casein and κ-casein. These peptides have proved active in the inhibition of ACE, a major factor in the regulation of blood pressure. Single oral administration of Calpis to SHR in a dose of 5 mL per kg of body weight was shown to significantly decrease systolic blood pressure between 4 and 8 h after administration. In a long-term feeding trial, a dose-dependent effect of Calpis administration to SHR was observed (*41*). Also, a preventive effect on hypertension was noted. In another study, the blood pressure-lowering effect of the two synthetic tripeptides was confirmed both in SHR and normotensive Wistar Kyoto rats (*42*). Both peptides showed dose dependent activity up to the dosage of 5 mg per kg of body weight. No similar effect was demonstrated in normotensive rats.

Hata et al. (*43*) carried out a placebo controlled study with hypertensive human subjects, who orally ingested a daily dose of 95 mL of Calpis sour milk for 8 weeks. Blood pressure was measured every 4 weeks. At 4 and 8 weeks of the trial both the systolic and diastolic blood pressure of the test group decreased significantly from the initial values. In the placebo group, no major changes in the blood pressure were observed.

Table IV. *In vivo* Studies about Bioactive Peptides

Subject	Product administered orally	Peptides identified/effect observed	Site of identification	Ref
Animal studies				
Minipig	Casein	β-casomorphins	Small intestine content	56
	Casein	CPP[b] α_{s1} (f 66-74)	Small intestine content	36
Calf	Skin milk	β-casomorphin	Blood plasma	37
	Skin milk	β-casomorphin CPP[b] immunostimulatory	Small intestine content	38
Rat	Casein	CPP[b]	Intestine content	57
	Lactoferrin	Lactoferricin	Intestine content	58
SHR[a]	Sour milk	Reduction of blood pressure	—	42
	Sour milk	Antihypertensive (Val-Pro-Pro-Ile-Pro-Pro)	Aorta	40
Human studies				
Adults	Milk	β-casomorphin	Small intestine content	35
New-born infants	Milk formula/ Human milk	Antithrombotic	Blood plasma	39
Adults	Sour milk	Reduction of blood pressure	—	43

[a] SHR = spontaneously hypertensive rat

[b] CPP = caseinophosphopeptide

Our recent studies (*44*) have shown that subcutaneous administration of a synthetic tetrapeptide, α-lactorphin (Tyr-Gly-Leu-Phe), lowered the blood pressure of SHR dose-dependently. The antihypertensive effector mechanism of this peptide was, however, not related to ACE-inhibition. α-lactorphin can be released from α-lactalbumin by pepsin-trypsin treatment and is known to exert a weak opioid activity (*59*).

Conclusions

An increasing number of *in vitro* and *in vivo* studies reveal that biologically active peptides are released from bovine milk proteins upon microbial fermentation and hydrolysis by digestive enzymes. Peptides with different bioactivities can be found in fermented milks and cheese varieties, but the 'specificity and amount of peptides formed is regulated by the starter cultures used and the rate of proteolysis. Highly proteolytic starter cultures, e.g., *Lactobacillus helveticus*, seem to produce short peptides with various bioactivities. On the other hand, weakly proteolytic dairy starter cultures, used regularly in the manufacture of fermented milk products, may not produce detectable amounts of bioactive peptides in the product but could be released *in vivo* after ingestion of the product. This phenomenon has been demonstrated under *in vitro* conditions, but remains to be shown *in vivo*. Also, the role of proteinases and peptidases isolated from different proteolytic bacteria needs to be elucidated with regard to the possibilities of producing specific peptides on a large scale, e.g., using continuous microbial fermentation. Of special interest in this context are the proteolytic properties of probiotic microorganisms. It is probable that part of the beneficial effects attributed to the foods containing probiotic bacteria is associated with bioactive peptides formed by the enzymes which are produced by these bacteria.

Some bioactive peptides have proved active in animal model and human studies. In this respect, the best experimental data is available with regard to antihypertensive peptides which inhibit the ACE system. Further studies are needed to establish the *in vivo* efficacy of other peptides as well as their long-term effects when administered orally on a regular basis. The bioactive milk peptides provide a highly interesting source to be used as active ingredients for the formulation of functional health-promoting foods.

References

1. Korhonen, H.; Pihlanto-Leppälä, A.; Rantamäki, P.; Tupasela T. *Trends Food Sci. Technol.* **1998**, *9*, 307-319.
2. Yamamoto, M. *Biopolymers* **1997**, *43*, 129-134.
3. Meisel, H. *Int. Dairy J.* **1998**, *8*, 363-373.
4. Xu, R. J. *Food Rev. Int.* **1998**, *14*, 1-16.
5. Marshall, V. W.; Tamime, A. Y. *Int. J. Dairy Technol.* **1997**, *50*, 35-41.
6. Meisel, H.; Bockelmann, W. *Antonie Van Leeuwenhoek.* **1999**, *76*, 207-215.
7. Poolman, B.; Kunji, E. R. S.; Hagting, A.; Juillard, V.; Konings, W. N. *J. Appl. Bact.* **1995**, *79*, 65S-75S.
8. Kunji, E. R.; Mierau, I.; Hagting, A.; Poolman, B.; Konings, W. N. *Antonie Van Leeuwenhoek* **1996**, *70*, 187-221.
9. Law, J.; Haandrikman, A. *Int. Dairy J.* **1997**, *7*, 1-11.
10. Konings, W. N.; Kuipers, O. P.; Huis in 't Veld, J. H. J. *Antonie Van Leeuwenhoek* **1999**, *76*, 1-4.
11. Konings, D. A.; Wyatt, J. R.; Ecker, D. J.; Freier, S. M. *J. Med. Chem.* **1997**, *40*, 4386-95.
12. Mierau, I.; Kunji, E. R. S.; Venema, G.; Kok, J. *Biotech. Genetic Engineering Rev.* **1997**, *14*, 279-301.
13. Palva, A. *Agric. Food Sci. Finland* **1998**, *7*, 267-282.
14. Hamel, U.; Kielwein, G.; Teschemacher, H. *J. Dairy Res.* **1985**, *52*, 139-48.
15. Matar, C.; Goulet, J. *Int. Dairy J.* **1996**, *6*, 383-397.
16. Pihlanto-Leppälä, A.; Rokka, T.; Korhonen H. *Int. Dairy J.* **1998**, *8*, 325-331.
17. Belem, M. A. F.; Gibbs, B. F.; Lee, B. H. *J. Dairy Sci.* **1999**, *82*, 486-493.
18. Yamamoto, N.; Akino, A.; Takano, T. *J. Dairy Sci.* **1994**, *77*, 917-922.
19. Maeno, M.; Yamamoto, N.; Takano, T. *J. Dairy Sci.* **1996**, *79*, 1316-1321.
20. Abubakar, A.; Saito, T.; Kitazawa, H.; Kawai, Y; Itoh, T. *J. Dairy Sci.* **1998**, *81*, 3131-3138.
21. Sütas, Y.; Hurme, M.; Isolauri, E. *Scand. J. Immunol.* **1996**, *43*, 687-689.
22. Sütas, Y.; Soppi, E.; Korhonen, H.; Syväoja, E-L.; Saxelin, M.; Rokka, T.; Isolauri, E. *J. Allergy Clin. Immunol.* **1996**, *98*, 216-224.
23. Majamaa, H.; Isolauri, E. *J. Allergy Clin. Immunol.* **1997**, *99*, 179-185.
24. Laffineur, E.; Genetet, N.; Leonil, J. *J. Dairy Sci.* **1996**, *79*, 2112-2120.
25. Kahala, M.; Pahkala, E.; Pihlanto-Leppälä, A. *Agric. Sci. Finl.* **1993**, *2*, 379-386.
26. Roudot-Algaron, F.; LeBars, D.; Kerhoas, L.; Einhorn, J.; Gripon, J. C. *J. Food Sci.* **1994**, *59*, 544-547.

27. Singh, T. K.; Fox, P. F.; Healy, A. *J. Dairy Res.* **1997**, *64*, 433-443.
28. Matar, C.; Amiot, J.; Savoie, L.; Goulet, J. *J. Dairy Sci.* **1996**, *79*, 971-979.
29. Addeo, F.; Chianes, L.; Salzano, A.; Sacchi, R.; Cappuccio, U.; Ferranti, P.; Malorni, A. *J. Dairy Res.* **1992**, *59*, 401-411.
30. Muehlenkamp, M. R.; Warthesen, J. *J. J. Dairy Sci.* **1996**, *79*, 20-26.
31. Rokka, T.; Syväoja, E-L.; Tuominen, J.; Korhonen, H. *Milchwissenschaft* **1997**, *52*, 675-678.
32. Moineau, S.; Goulet, J. *Milchwissenschaft* **1991**, *46*, 551-553.
33. Meisel, H.; Goepfert, A.; Gunther, S. *Milchwissenschaft* **1997**, *52*, 307-311.
34. Ryhänen, E-L.; Pihlanto-Leppälä, A.; Pahkala, E. *Int. Dairy J.* **2000**, submitted.
35. Svedberg, J.; de Haas, J.; Leimenstoll, G.; Paul, F.; Teschemacher, H. *Peptides* **1985**, *6*, 825-830.
36. Meisel, H.; Frister, H. *J. Dairy Res.* **1989**, *56*, 343-349.
37. Umbach, M.; Tecshemacher, H.; Praetorius, K.; Hirchhäuser, H.; Bostedt, H. *Reg. Peptides* **1988**, *12*, 223-230.
38. Scanff, P.; Yvon, M.; Thirouin, S.; Pélissier, J-P. *J. Dairy Res.* **1992**, *59*, 437-447.
39. Chabance, B.; Jollés, P.; Izquierdo, C.; Mazoyer, E.; Francoual, C.; Drouet, L.; Fiat, A-M. *British J. Nutr.* **1995**, *73*, 583-590.
40. Masuda, O.; Nakamura, Y.; Takano, T. *J. Nutr.* **1996**, *126*, 3063-3068.
41. Nakamura, Y.; Yamamoto, N.; Sakai, K.; Okubo, A.; Yamazaki, S.; Takano, T. *J. Dairy Sci.* **1995**, *78*, 777-783.
42. Nakamura, Y.; Yamamoto, N.; Sakai, K.; Takano, T. *J. Dairy Sci.* **1995**, *78*, 1253-1257.
43. Hata, Y.; Yamamoto, M.; Ohni, H.; Nakajima, K.; Nakamura, Y.; Takano. T. *Amer. J. Clin. Nutr.* **1996**, *64*, 767-771.
44. Nurminen, M.-L.; Sipola, M.; Kaarto, H.: Pihlanto-Leppälä, A.; Piilola, K.; Korpela, R.; Tossavainen, O.; Korhonen, H.; Vapaatalo, H. *Life Sci.* **2000**, *66*, 1535-1543.
45. Mellander, O. *Acta Society Medicine Uppsala* **1950**, *55*, 247-255.
46. Brantl, V.; Teschemacher, H.; Henschen, A.; Lottspeich, F. *Hoppe-Seyler's Z. Physiol. Chem.* **1979**, *360*, 1211-1216.
47. Zioudrou, C.; Streaty, R. A.; Klee, W. A. *J. Biol. Chem.* **1979**, *254*, 2446-2449.
48. Jolles, P.; Parker, F.; Floc'h, F.; Migliore, D.; Alliel, P.; Zerial, A.; Werner, G. G. *J. Pharmacol.* **1981**, *3*, 363-369.
49. Maruyama, S.; Suzuki, H. *Agric. Biol. Chem.* **1982**, *46*, 1393-1394.

186

50. Yoshikawa, M.; Tani, F.; Ashikaga, T.; Yoshimura, T.; Chiba, H. *Agric. Biol. Chem.* **1986**, *50*, 2951-2954.
51. Chiba, H.; Yoshikawa, M. In *Protein Tailoring for Food and Medical Uses,* Feeney, R. E.; Whitaker, J. R., Eds.; Marcel Dekker: New York, **1986**, pp. 123-153.
52. Jolles, P.; Lévy-Toledano, S.; Fiat, A.-M.; Soria, G.; Gillessen, D.; Thomaidis, A.; Dunn, F. W.; Caen, J. P. *Eur. J. Biochem.* **1986**, *158*, 379-382.
53. Tomita M.; Bellamy, W.; Takase, M.; Yamauchi, K.; Wakabayashi, H.; Kawase, K. *J. Dairy Sci.* **1991**, *74*, 4137-4142.
54. Mullally, M. M.; Meisel, H.; FitzGerald, R. J. *Biol. Chem. Hoppe-Seyler* **1996**, *377*, 259-260.
55. Smacchi, E.; Gobbetti, M. *Enzyme Microb. Technol.* **1998**, *22*, 687-694.
56. Meisel, H. *FEBS Lett.* **1986**, *196*, 223-227.
57. Sato, R.; Naguchi, T.; Naito, H. *J. Nutr. Sci. and Vitaminol.* **1986**, *32*, 67-76.
58. Tomita, M.; Takase, M.; Bellamy, W.; Shimamura, S. *Acta Paediatr. Jpn.* **1994**, *36*, 585-591.
59. Antila, P.; Paakkari, I.; Järvinen, A.; Mattila, M. J.; Laukkanen, M.; Pihlanto-Leppälä, A.; Mäntsälä, P.; Hellman, J. *Int. Dairy J.* **1991**, *1*, 215-229.

Chapter 14

Effects of Enzyme Hydrolysis and Thermal Treatment on Bioactive Flavor Compounds in Pork-Based Flavoring Ingredients

Mathias K. Sucan[1], Erica A. Byerly[2], Ingolf U. Grün[2], Lakdas N. Fernando[2], and Nayan B. Trivedi[1]

[1]Applied Food Biotechnology, Inc., 937 Lone Star Drive, O'Fallon, MO 63366
[2]Department of Food Science, University of Missouri at Columbia, 256 William C. Stringer Wing, Columbia, MO 65211

Pork loin tissues were hydrolyzed by either one of two endopeptidases, alcalase or papain, under optimal condition of 60°C, 60 min. The hydrolysates were subjected further to heat treatment at 90°C, 30 min or 120°C, 30 min. The resulting pork-based flavoring ingredients were subsequently evaluated for organoleptic properties using a descriptive panel. Sensory scores indicated that chicken aroma was the most predominant aroma attribute in all pork-based flavoring ingredients obtained from both enzyme digestion processes and temperature treatments. The chicken aroma attribute varied indiscriminately of enzyme type and thermal processing conditions. Flavor volatiles were obtained by GC-MS analysis to elucidate the unexpected development of intense chicken aroma note. Lipid-derived aldehydes, implicated in the development of this note, were affected by both temperature and enzyme treatments. Proposed chemical and biochemical mechanisms involved in the development of chicken flavor from pork-based flavoring ingredients are discussed.

Raw meat has only a weak sweet aroma resembling serum, and a salty, metallic, bloody taste. However, it is a rich reservoir of compounds with taste properties and aroma precursors (*1*). Each meat has a distinct flavor characteristic. The flavors of distinct meat species or species-specific flavors are often carried by the lipid fraction (*2*). For example 4-methyloctanoic and 4-methyldecanoic acids are specific to mutton while (*E,E*)-2,4-decadienal is specific to poultry meat. 12-Methyltridecanal has been identified as species-specific odorant in stewed beef, and is responsible for the tallowy and beef-like smell (*3*). The distinctive pork-like or piggy flavor noticeable in lard has in part been attributed to *p*-cresol and isovaleric acid (*4,5,6*).

The first step in making meat-based flavoring ingredients can involve enzymatic digestion. Papain and alcalase, are industrially used to digest muscle proteins. Papain, produced from papaya latex, is an endopeptidase. It is also a cysteine protease, in which cysteine is an essential part of the catalytic site. This enzyme is traditionally used in meat tenderizing (*7*). Alcalase is also an endopeptidase, but is a serine protease produced from the bacterial species *Bacillus licheniformis*. It is one of the best enzymes for producing soluble protein hydrolysates from a number of protein sources. It has a broad specificity and cleaves many types of peptide bonds, preferentially those with a hydrophobic side chain on the carbonyl side (*7*).

Non-volatile precursors of meat aromas are formed upon hydrolysis of meat substrate biopolymers, and include peptides, amino acids, nucleotides, vitamins, and reducing sugars. These compounds contribute to sweet, salty, sour, bitter and umami sensations (see Figure 1). While the sugars contribute to sweet taste, and the acid to sour taste exclusively, the amino acids and simple peptides elicit all the 5 primary taste sensations. The umami compounds have no characteristic flavor of their own at the concentrations in which they occur in foods. Rather, they have the ability to modify taste through their flavor enhancing capability, and their ability to act synergistically when used in combination with each other. Most of the aroma-active compounds in meat are generated upon heating from non-volatile precursors as an effect of lipid oxidation and degradation, and from degradation and interactions of sugars, amino acids, ribonucleotides, proteins, pigments and vitamins (Figure 2). Approximately 1000 flavor compounds have been identified in meat from beef, poultry, pork and sheep. These compounds range from the simple hydrocabons, aldehydes, ketones, alcohols, carboxylic acids, esters, ethers, to the more complex lactones, furans, pyrroles, pyridines, pyrazines, thiophenes, thiazoles, oxazoles, and other sulfur and nitrogen-containing substances. In spite of the high number of flavor compounds in meats, only a small fraction possesses meat aroma characteristics (*8*), and are sulfur containing in nature.

In poultry meat, Maillard reactions and lipid oxidation are major sources of volatile flavor compounds (*9*). The meaty flavor of chicken broth has been found to be due to 2-methyl-3-furanthiol (*10*). This compound has also been recognized as a character impact compound in the aroma of cooked beef (*9,10*). However, a major difference between beef and chicken is that bis(2-methyl-3-furyl) disulfide, the oxidation product of 2-methyl-3-furanthiol and responsible

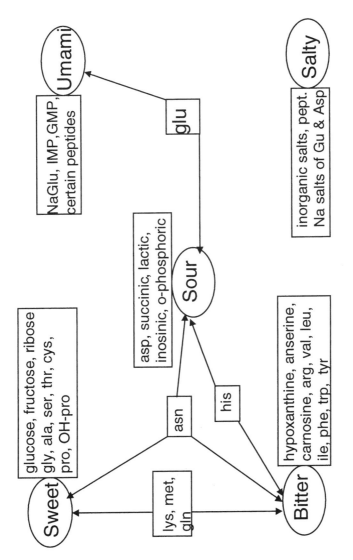

Figure 1. Taste active compounds.

190

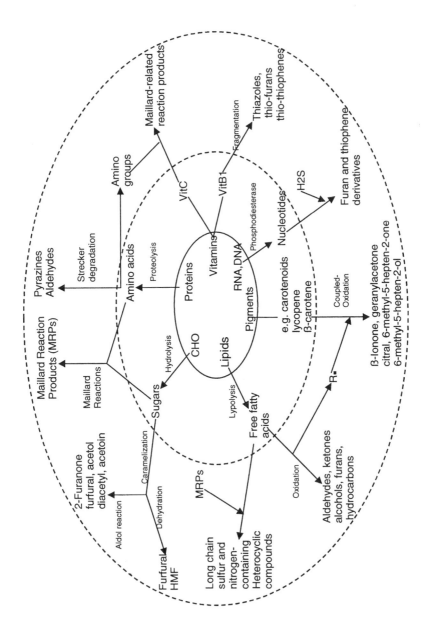

Figure 2. Wheel of Chemical Reactions Important to Flavor Formation

for a meaty note, and methional which is responsible for a cooked potato attribute, predominate in beef whereas volatiles from the oxidation of unsaturated lipids, in particular E,E-2,4-decadienal possessing a fatty taste, and γ-dodecalactone possessing a tallowy, fruity taste, prevail in chicken (*9,10*). Schroll and co-workers (*11*) found that the most abundant aldehyde in chicken flavor are hexanal and 2,4-decadienal, and that the enzymatic hydrolysis of chicken with papain increased the concentration of 2,4-decadienal from 5.2 to 13.7 mg/Kg.

Lipid-derived volatile compounds dominate the flavor profile of pork cooked at temperatures below 100°C. The large numbers of heterocyclic compounds reported in the aroma volatiles of pork are associated with roasted meat rather than boiled meat where the temperature does not exceed 100°C (*12,13*). Of the volatiles produced by lipid oxidation, aldehydes are the most significant flavor compounds (*13,14*). Octanal, nonanal, and 2-undecenal are oxidation products from oleic acid, and hexanal, 2-nonenal, and 2,4-decadienal are major volatile oxidation products of linoleic acid. Interestingly enough, oleic acid and linoleic acid are the two most abundant unsaturated fatty acids in pork (*13,15*).

From the above information, one would expect a meat-based flavoring ingredient to elicit flavor attributes reminiscent of the odor of the animal species that served as meat source. In a recent in-house investigation on meat substrates for their impact on the quality of savory flavoring ingredients, however, pork-based flavoring ingredients unexpectedly developed intense chicken aroma. To elucidate some fundamental bases underlying this observation, a GC/MS study was undertaken.

Experimental Studies

Two enzyme systems for digestion, and reaction conditions were used to process meat substrates. The resulting products were evaluated for human sensory attributes and volatile flavor composition.

Sample Procurement

USDA grade chicken breast and pork loin were obtained from a local grocery store. Alcalase was obtained from Novo Nordisk (Franklinton, NC) and papain from Enzyme Development Company (New York, NY).

Sample Preparation

Each meat substrate was ground by means of a Hobart commercial food preparing machine (Hobart Co., Troy, Ohio) equipped with a 12 5/64 plate. Ground meat from each source was mixed with about equal weight of deionized water to make slurry, which was subsequently subjected to enzymatic hydrolysis and thermal processing.

Enzymatic digestion: Two proteases, papain and alcalase, were used to digest muscle proteins in pork and chicken breast. It was important to determine the adequate digestion conditions for each meat substrate by monitoring degree of hydrolysis (DH) and viscosity. Appropriate amounts of slurries were hydrolyzed with 0.1% enzyme at 60°C. Aliquots of hydrolysates were taken every 15 min for viscosity and DH measurements over a 90 minute period. A digestion time of 1 hour was found adequate for both enzyme systems and both meat substrates. DH determination was based on formol titration method (*16*) which is similar to the pH-stat technique.

Thermal processing: After hydrolysis, 2% glucose was added to the slurries, and this was followed by reaction at 90°C for 30 min or 120°C for 30 min. The resulting product was stabilized by addition of potassium sorbate, benzoic acid and mixed tocopherols. The pH of the products was adjusted to approximately 4.2 using phosphoric acid.

Descriptive sensory analysis

A sensory panel was trained to utilize a 15 cm unstructured line scale to evaluate each flavoring attribute's intensity from low to high. The terms that were used for aroma were pungent, meaty, chicken, pork, brothy, musty, burnt, and roasted. Judges' marks were measured, and then statistical analysis was performed on the data to provide detailed information about the product's attributes, and how well the judges scored them. Analysis of variance (ANOVA) was used to see whether treatment variables such as processing conditions, and meat substrates had an effect on the sensory properties of products. ANOVA provides a way of comparing several means at the same time, and estimates the variance attributed to each factor (in this case, variance attributed to Judge, Replicate, Product, Judge to Replicate, Judge to Product, and Replicate to Product interactions).

GC/MS Study of Volatile Flavors

5 Gram sample was placed into a round-bottom flask and mixed with 30 mL HPLC grade water. The flask was fitted to a condenser and a Tenax (60/80 mesh) adsorption tube. The flask was placed into a sandbath at 80C (with sample temperature at 70°C) and nitrogen at a flow rate of 30 mL/min was used to sweep the headspace for 30 min. Volatiles were desorbed from the Tenax by thermal desorption in a Dynatherm ACEM900 thermal desorption unit (Dynatherm Analytical Instruments, Kelton, PA) with desorption, valve, trap and transfer line temperatures of 250°C, 200°C, 250°C and 250°C, respectively. Separation of volatiles was accomplished on a 60 m DB-5MS capillary column (0.25 mm id, 0.25 um film) (J&W Scientific, Folsome, CA) with helium flow rate of 0.65 mL/min using a Varian GC 3400CX. The temperature program was 35°C for 5 min, then increased at 8°C/min to 220°C and then to 250°C at 2 C/min. A Varian Saturn 2000 MS (Varian, Walnut Creek, CA) was used in the EI mode with 70 ev and a 5 min filament delay. The ion trap temperature was 150°C and masses were scanned from 40-350 m/z. NIST92 and Wiley5 libraries of mass spectra and comparison of retention indices to published data, as well as selected standards were used for tentative identification of compounds.

Results and Discussion

Human Sensory Evaluation

To test the validity of sensory experiment, ANOVA was performed for the judges' scores for aroma, and interactions of Judge with Replicate, Judge with Product, and Replicate with Product. The analysis provided detailed information about the entire panel, individual judges, replicates, products, and their interactions. A probability value (p = 0.0001) indicates that the ANOVA model was significant and adequately fit the sensory data. The variable Replicate was not significant (p=0.24), indicating that replicates were not different. Products were significantly different (p=0.0001), which makes sense since there were different meat substrates. There was no significant interaction (p=0.16) between Judge and Replicate, while Judge and Product interaction was significant (p=0.0001). Interaction between Replicate and Product was not significant (p=0.52), indicating that the rating of Replicate was not influenced by Product. The p-values for meaty, chicken, pork, brothy, musty, burnt, and roasted were larger than 0.05 for the interactions between replicate and product, indicating that the products were scored similarly between replicates for these attributes. Overall, ANOVA indicated that the attribute "chicken aroma" was adequately evaluated by panelists.

The mean sensory scores for pungency, meaty, chicken, brothy, musty, burnt and roasted attributes are given in Table I. Scores indicate that the "chicken" attribute was the most intense note in all flavoring ingredients, and the least perceived flavor descriptor was the burnt note. The flavoring ingredients were made from different meat substrates, enzymes and thermal treatments. The type of meat substrate and temperature treatment had an effect on sensory scores.

CBPA and CBAA were more pungent, PPB, PAB and CBAB more meaty, CBPA, CBPB, PAA, CBAA, and CBAB had the strongest chicken aromas, PPB, CBPA, PPA, and CBPB were more brothy, PAA, CBAA and CBPA more musty, PPB, PAB CBPB and CBPA smelled roasted. Pungency was more noticed in chicken breast at lower temperature. At the higher temperature treatment, the compounds (more likely lipid-related aldehydes) responsible for this sensation diminished certainly due to loss by volatilization and/or further oxidation reactions or reaction with hydrogen sulfide (H_2S). Meatiness was more characteristic of products treated at higher temperature as a result of more pronounced nonenzymatic and degradation reactions. Lower temperature products were mustier, with mustiness decreasing with increasing temperature. The development of brothiness did not follow any specific pattern. The higher temperature-treated products were about four times more roasted, certainly as a result of Maillard reactions. The burnt note was weak and not characteristic of any particular product or process. Chicken aroma developed in products irrespective of temperature treatment and enzyme systems. To our surprise, pork meat developed intense chicken aroma note upon processing. The biochemical and chemical bases involved in such a development were further investigated by GC/MS analysis.

Aroma Evaluation by GC/MS Analysis

This study identified several aldehydes, ketones, alcohols, hydrocarbons, heterocyclics and an alkylsulfide in all flavoring ingredients independently of the kind of meat substrate, the enzyme used in digestion, and the temperature treatment (see Table II). From a qualitative standpoint there was a striking resemblance between the aroma profile of chicken-based and that of pork-based flavoring ingredients, except for 2-methylpropanal, 2-heptanone, 2-undecanone, octanol and pentanol in chicken-based ingredients and 1-octen-3-one in pork-based ingredients. Furthermore, lipid oxidation products dominated the volatile flavor profiles. Mottran et al. (2) reported that boiled pork volatile flavors were dominated by aliphatic aldehydes and alcohols with only a very small amount of heterocyclic compounds. This study identified only few heterocyclic compounds; two furans and a thiophene. This may be due to the relatively higher formation of carbonyl compounds. Lipids reduce the formation of heterocyclic compounds by Maillard reaction. Carbonyl compounds derived from the autoxidation of lipids capture reactants such as H_2S essential for the formation of heterocyclic compounds (*17*).

Of the compounds isolated from the flavoring ingredients, which ones contributed to the chicken aroma characteristic? Table III lists sets of flavor chemicals reported to elicit chicken aromas. Using aroma extract dilution analysis (AEDA) and simultaneous distillation extraction (SDE) techniques, Gasser and Grosch (*10*) found 16 primary odorants in chicken broth. Fourteen of these were identified as 2-undecenal, *E*-2-nonenal, *E,E*-2,4-nonadienal, *E,E*-2,4-decadienal, methional, nonanal, γ-decalactone, γ-dodecalactone, β-ionone, *p*-cresol, 2-methyl-3-furanthiol, 2-formyl-5-methylthiophene, 2,4,5-trimethylthiazole and 2-furfurylthiol. More recently, Kerler and Grosch (*18*) reported *E,Z*- and *E,E*-2,4-decadienal, *E,E*-2,4-nonadienal, 2-furfurylthiol, hexanal, octanal and acetaldehyde to be the character impact compound of freshly boiled chicken. During the same year, Farkas et al. (*19*) reported that the most potent odorants of pressured-cooked hen meat were 2-furfurylthiol, 3-(methylthio) propanal, 2,5-dimethyl-4-hydroxy-3(2H)-furanone, *E,E*-2,4-decadienal, 2-methyl-3-furanthiol, 2-ethyl-3,5-dimethyl pyrazine, 2,4,6-trimethyltetrahydro-1,3,5-thiadizine, 3,5-dimethyl-1,2,4-trithiolane and 5,6-dihydro-2,4,6-trimethyl-4H-1,3,5-dithiazine. Of all these flavor chemicals found to carry the flavor of chicken, the present study identified only hexanal, octanal, nonanal, *E,E*-2,4-decadienal, *E*-2-nonenal and methional. Because different studies reported different set of compounds carrying the flavor of boiled chicken or chicken broth (see Table III), it appears that several sets of flavor chemicals may be responsible for the chicken aroma note, and no single combination of chemicals can claim responsibility for this sensory attribute. Since all the chicken aroma-eliciting compounds we identified are lipid-derived aldehydes, discussion on effects of processing on flavor quality will focus on these lipid oxidation products.

Meat origin, enzyme type and thermal treatment had an impact on the qualitative and quantitative aroma profiles of meat-based flavoring ingredients. From a matrix comparison standpoint, pork-made flavorings had, in most cases, greater concentrations of volatiles than chicken-made products (Table II). Lipid-derived volatiles predominated in pork-based flavorings. Heterocyclic compounds were the second most abundant volatile flavor compounds identified. About twice as much 2-methylfuran was found in pork-based compared with chicken-based products, while twice as much 2-pentylfuran was found in chicken-based compared with pork-based flavorings. Interestingly, linoleic acid is a precursor for 2-pentylfuran, and chicken meat contains twice as much linoleates as pork meat. It is also not surprising that octanal and nonanal, which derive from the oxidation of oleic acid, dominated the spectrum of lipid-related volatiles, with twice as much volatiles in pork as in chicken; oleic acid is the most abundant fatty acid in pork fat (~44%) and in chicken fat (~38%). Sulfur compounds were found in appreciable amounts, especially in pork-based ingredients.

Table I. Aroma Attributes of Flavoring Ingredients

	Pungent	Meaty	Chicken	Brothy	Musty	Burnt	Roasted
PPA	1.7^{BCD}	0.8^{D}	3.9^{BC}	2.3^{C}	2.8^{BCD}	0.08^{BC}	0.8^{E}
PPB	1.2^{D}	3.8^{A}	3.3^{C}	4.3^{A}	2.7^{BCD}	0.3^{ABC}	3.7^{A}
CBPA	2.2^{ABCD}	0.8^{D}	5.5^{A}	3.5^{ABC}	3.2^{AB}	0.4^{ABC}	0.6^{E}
CBPB	1.5^{CD}	2.3^{BC}	6.1^{A}	3.2^{ABC}	2.2^{CD}	0.6^{ABC}	2.8^{AB}
PAA	1.3^{D}	1.2^{CD}	5.7^{A}	3.5^{ABC}	3.5^{AB}	0.2^{BC}	1.0^{DE}
PAB	1.5^{CD}	3.3^{AB}	3.9^{BC}	2.5^{C}	2.7^{CD}	0.4^{ABC}	2.4^{BC}
CBAA	1.9^{ABCD}	0.8^{D}	5.6^{A}	3.0^{BC}	3.5^{AB}	0.03^{C}	0.7^{E}
CBAB	1.2^{D}	2.6^{AB}	5.3^{AB}	2.7^{BC}	1.9^{D}	0.5^{ABC}	1.9^{CD}

The letters in the superscript indicate the intensity of a product's aroma attribute, with the A's being of high intensity, B of less intensity, and so forth.

Means within columns followed by the same superscript letters do not differ significantly ($p < 0.05$).

PPA- Pork digested with Papain, reacted at 90°C, 30 min.
PPB- Pork digested with Papain, reacted at 120°C, 30 min.
PAA- Pork digested with Alcalase, reacted at 90°C, 30 min.
PAB- Pork digested with Alcalase, reacted at 120°C, 30 min.
CBPA- Chicken Breast digested with Papain, reacted at 90°C, 30 min.
CBPB- Chicken Breast digested with Papain, reacted at 120°C, 30 min.
CBAA- Chicken Breast digested with Alcalase, reacted at 90°C, 30 min.
CBAB- Chicken Breast digested with Alcalase, reacted at 120°C, 30 min.

Table II. Volatile flavor composition of flavoring ingredients

Compounds	Papain-treated Chicken-based Flavors		Papain-treated Pork-based Flavors	
	90C, 30 min. Conc., ppb	120C, 30 min. Conc., ppb	90C, 30 min. Conc., ppb	120C, 30 min. Conc., ppb
ALDEHYDES				
2-methylpropanal	15.5	49	28	185
pentanal	19.2	160	11	97
2-methyl-2-butenal	9	7	74	111
hexanal	57	140	6	14
heptanal	24	19	36	59
benzaldehyde	14	37	67	20
octanal	27	22	78	26
nonanal	32	7	16.5	10
E,E-2,4-decadienal	19	11	33	10
E-2-nonenal	11	8.4		
methional		17		10
KETONES				
2-undecanone	13	8		
1-octen-3-one			4	8
ALCOHOLS				
1-octanol	27	110		
3-methyl cyclohexanol	48	73	11	44
1-octen-3-ol	31	34	28	23

HYDROCARBONS				
toluene	1	13	3	11
dimethyl benzene	5	4		
p-xylene	5	6	70	56
trimethyl benzene		10	13	15
dodecene	1.2	36	20	14
tridecene	2	2	15	7
tetradecene		7	26	11
HETEROCYCLICS				
2-methylfuran	81	137	223	309
3-methylthiophene	28	17	23	10
2-pentylfuran	32	214	12	100
SULFUR-COMPOUNDS				
dimethyl disulfide	11	13	33	70.5

Continued on next page.

Table II. Continued

Compounds	Alcalase-treated Chicken-based Flavors		Alcalase-treated Pork-based Flavors	
	90C, 30 min. Conc., ppb	120C, 30 min. Conc., ppb	90C, 30 min. Conc., ppb	120C, 30 min. Conc., ppb
ALDEHYDES				
2-methylpropanal	3.1	5		
pentanal	23	27	26	154
2-methyl-2-butenal	9	8.95	9	9
hexanal	48.2	117	231	340
benzaldehyde	15	21	35	51
octanal	51	9	38	72
nonanal	88	14	30	65
E,E-2,4-decadienal	66	7	29	90
E-2-nonenal	49	7	33	14
methional	1.2	12	4	1.7
KETONES				
2-heptanone	7			
1-octen-3-one		14	4	5
ALCOHOLS				
1-pentanol	14	10		
3-methyl				
cyclohexanol	95	113	54	154
1-octen-3-ol	27	70	12	30

HYDROCARBONS				
toluene	13	30	14	28
methyl cyclopentane	2	60		
dimethyl benzene			1	1
p-xylene			7	9
trimethyl benzene	5	47	12	16
dodecene	16.9	20	26	54
tridecene			11.3	26
tetradecene			15	28
HETEROCYCLICS				
2-methylfuran	86	72	203	144
3-methylthiophene	22	22	23	14
2-pentylfuran	113	497	29	288
SULFUR-COMPOUNDS				
dimethyl disulfide	17	45	39	74

Table III. Sets of Compounds Reported to Carry Chicken Flavor

	Chicken broth (Gasser & Grosch, 1990) FD Factors	Boiled chicken (Kerler & Grosch, 1997)	Pressure cooked chicken (Farkas et al., 1997)
2-undecenal	256		
E-2-Nonenal	64		
E,E-2,4-nonadienal	64	+	
E,E-2,4-decadienal	2058	+	+
methional	128		
nonanal	128		
γ-decalactone	64		
γ-dodecalactone	128		
β-ionone	512		
2,5-dimethyl-3-furanthiol	256		
2-methyl-3-furanthiol	1024		+
2-formyl-5-methylthiophene	64		
2,4,5-trimethylthiazole	128		
2-furfurylthiol	512	+	+
hexanal		+	
octanal		+	
acetaldehyde		+	
E,Z-2,4-Decadienal		+	
3-(methylthio) propanal			+
furaneol			+
2-ethyl-3,5-dimethyl pyrazine			+

The amount of lipid-derived volatiles increased with increasing temperature from 90°C to 120°C, for pentanal, 2-methyl-2-butenal, hexanal, heptanal and benzaldehyde, while octanal, nonanal, *E,E*-2,4-decadienal and *E*-2-Nonenal decreased in concentrations, for papain-digested pork. With alcalase-digested pork flavorings, the decrease was only observed in the concentrations of *E,E*-2,4-decadienal and *E*-2-nonenal. The very high concentration of hexanal suggests that *E,E*-2,4-decadienal broke down into hexanal, with *E*-2-nonenal as an intermediate. This is confirmed by the chemical pathway of hexanal formation from linoleic acid proposed by Shi and Ho (*9*).

In this study, hexanal was the dominant aldehyde in both papain and alcalase-treated products. This confirms a previous study by Schroll and co-workers (*11*) who found that the most abundant aldehydes in papain hydrolysates of chicken were hexanal and 2,4-decadienal. With respect to enzyme types, flavor concentrations were generally the same in all products from lower temperature treatment. With higher temperature treatment, however, the products from alcalase digestion exhibited greater concentrations in flavor volatiles than those from papain digestion.

The unexpected development of chicken flavor in pork-based flavoring ingredients may have been du. to an extensive production of lipid-derived volatile compounds aided by protein hydrolysis. Schroll et al. (*11*) reported that the enzymatic hydrolysis of chicken with papain increased the concentration of 2,4-decadienal from 5.2 to 13.7 mg/Kg and improved the aroma of cooked meat. The enzymatic digestion of meat tissue proteins may expose more meat lipids to autoxidation processes. As a result, greater amounts of lipid-related volatiles are formed and may modify the overall flavor perception. In fact, the addition of sunflower oil changed the odor quality of beef broth to that of chicken broth (*20*). Gasser and Grosch (*10*) reported that carbonyl compounds formed by oxidative degradation of unsaturated acyl lipids carry the chicken aroma note since their removal from the volatile fraction resulted in loss of chicken odor, with intensification of the meaty odor. Cabonyl compounds formed by the autoxidation of unsaturated lipids, may have changed the pork meat-like odor to a chicken-like odor.

Conclusions

Analysis of variance indicated that the attribute "chicken aroma" was adequately evaluated by panelists and was the most important aroma in all pork-based flavoring ingredients. Chicken aroma developed in products irrespective of temperature treatment and enzyme systems.

Subsequent GC/MS analysis of flavorings identified several compounds of different chemical classes, but only hexanal, octanal, nonanal, *E,E*-2,4-decadienal, *E*-2-nonenal and methional, were previously reported to be carrier of chicken flavor. Because different studies reported different sets of compounds carrying the flavor of boiled chicken or chicken broth, it appears that several sets of flavor chemicals may be responsible for the chicken aroma note, and no single combination of chemicals can claim responsibility for this sensory attribute.

Meat origin, enzyme type and thermal treatment had an impact on the qualitative and quantitative aroma profiles of meat-based flavoring ingredients. From a matrix comparison standpoint, pork-made flavorings had, in most cases, greater concentrations of lipid-derived volatiles than chicken-made products. The amount of lipid-derived volatiles increased with increasing temperature from 90°C to 120°C, for pentanal, 2-methyl-2-butenal, hexanal, heptanal and benzaldehyde, while octanal, nonanal, *E,E*-2,4-decadienal and *E*-2-Nonenal decreased in concentrations. *E,E*-2,4-decadienal broke down into hexanal, using *E*-2-nonenal as intermediate, as indicated by the very high concentration of hexanal. With respect to enzyme types, flavor concentrations were generally the same in all products from lower temperature treatment. With higher temperature treatment, however, the products from alcalase digestion exhibited greater concentrations in flavor volatiles than those from papain digestion.

The unexpected development of chicken flavor in pork-based flavoring ingredients was more probably caused by an extensive production of lipid-derived volatile compounds, which may have changed the pork-like aroma to a chicken-like aroma in pork-based flavoring ingredients.

References

1. Shahidi F. In *Flavor of Meat and Meat Products*; Shahidi, F., ed.; Blackie Academic & Professional: New York, NY, 1994.
2. Mottram, D. S.; Edwards, R. A; McFie, H. J. *J. Sci. Food Agric.* **1982**, *33*, 934-944.
3. Guth, H; Grosch, W. *Lebensmittel-Wissenschaft-und-Technologie* **1993**, *26*, 171-177.
4. Ha, J.K.; Lindsay, R.C. *J. Food Sci.* **1991**, *56*, 1197-1202.
5. Ha, J.K.; Lindsay, R.C. *Lebensmittel-Wissenschaft und –Technologie* **1990**, *23*, 433-440.
6. Lindsay, R.C. In *Food Chemistry*; O. R. Fennema, ed.; Marcel Dekker: New York, 1996: pp 723-765.
7. Adler-Nissen, J. In *Enzymes in Food Processing*; T. Nagodawithana; G. Reed, eds; Academic Press, Inc.; New York, NY, 1993.
8. Shahidi F. In *Thermal Generation of Aromas;* T. H. Parliament; R. J. McGorrin; C.-T. Ho, eds; ACS Symposium Series 409; American Chemical Society: Washington, DC, 1989.
9. Shi, H.; Ho, C.-T. In *Flavor of Meat and Meat Products*; F. Shahidi, ed.; Blackie Academic & Professional: New York, NY, 1994.

10. Gasser, U.; Grosch, W. *Z Lebensm Unters Forsch* **1990**, *190*, 3-8.

11. Schroll, W.; Siegfried, N.; Drawert, F. *Z Lebensm Unters Forsch* **1988**, *187*, 558-560.

12. Mottram, D. S. *J. Sci. Food Agric.* **1985**, *36*, 377-382.

13. Ho, C.-T.; Oh, Y.-C.; Bae-Lee, M. In *Flavor of Meat and Meat Products*; F. Shahidi, ed; Blackie Academic & Professional: New York, NY, 1994.

14. Frankel, E. N. *Prog. Lipid Res.* **1982**, *22*, 1-33.

15. Schliemann, J.; Wolm, G.; Schrodter, R.; Ruttloff, H. *Nahrung* **1987**, *31*, 47-56.

16. Nielson, P. M.; Glenvig, H. Novo Nordisk: Franklinton, NC, 1997.

17. Whitfield, F. B.; Mottram, D. S.; Brock, S.; Puckey, D. J.; Salter, L. J. *J Sci. Food Agric* **1988**, *42*, 261-272.

18. Kerler, J.; Grosch, W. *Z Lebensm Unters Forsch* **1997**, *205*, 232-238.

19. Farkas, P.; Sadecka, J.; Kovac, M.; Siegmund, B.; Leitner, E.; Pfannhauser, W. *Food Chemistry* **1997**, *60*, 617-621.

20. Gasser, U.; Grosch, W. *Z Lebensm Unters Forsch* **1988**, *186*, 489-494.

Chapter 15

Effect of Nixtamalization on Fumonisin-Contaminated Corn for Production of Tortillas

Mary A. Dombrink-Kurtzman[1] and Lloyd W. Rooney[2]

[1]Mycotoxin Research Unit, National Center for Agricultural Utilization Research, Agricultural Research Service, U.S. Department of Agriculture, 1815 North University Street, Peoria, IL 61604
[2]Cereal Quality Laboratory, Department of Soil and Crop Sciences, Texas A&M University, College Station, TX 77843

Fumonisins, mycotoxins produced by *Fusarium verticilliodes* (Sacc.) Niremberg (synonym *F. moniliforme* Sheldon) and *Fusarium proliferatum*, are found in corn worldwide. Low levels of fumonisins can occur in corn products destined for human consumption. Studies were undertaken to determine the fate of fumonisins during nixtamalization (alkaline cooking), using normal-appearing corn that was naturally contaminated with fumonisin B_1 at 8.8 ppm. Samples from each stage of processing were analyzed to determine how much fumonisin remained in finished products. The majority of the fumonisin (76%) was present, primarily as hydrolyzed fumonisin B_1, in the steep water and wash water. Tortillas contained approximately 0.50 ppm fumonisin B_1, plus 0.36 ppm hydrolyzed fumonisin B_1, representing 18.5% of the fumonisin B_1 detected in the raw corn. Nixtamalization appears to be a means for significantly reducing the amount of fumonisin in corn.

In many parts of the world, corn (maize) is the basic food item. Tortillas have long been a staple food in both Mexico and Central America, where maize has been the traditional cereal for preparation of tortillas (*1*). In Mexico, the *per capita* consumption of maize is approximately 250 g/day, mainly as tortillas; in certain populations the intake is even higher (*2*). Increased popularity of tortillas and other Mexican food is occurring in the United States, Europe and Asia. The U. S. overall market for soft corn tortillas has been estimated at $2.87 billion in 1996 by the Tortilla Industry Association (*2*). In addition, more than $4.3 billion worth of tortilla and corn chips are produced each year in the United States, with yearly sales increases of 8-10% (*3*).

Nixtamalization (Alkaline Cooking)

Physical Changes

Corn, lime and water are the three basic ingredients required for nixtamalization to produce masa, which is then processed into tortillas. A flowchart of the traditional process for producing tortillas is shown in Figure 1. The specific corn hybrid, environmental conditions during growth and storage and handling procedures significantly affect the corn cooking time. The actual processes for masa production can differ because a variety of different conditions, some of which are proprietary, are used by commercial plants for cooking and steeping corn to produce nixtamal (steeped corn). The food industry, in general, prefers to process corn hybrids, which have pericarp (hull) that can be removed during cooking. Most of the dry matter loss in commercially processed corn is pericarp, which is removed, along with excess lime during the washing step. Removal of pericarp is an important means for reducing the amount of fumonisin present in the finished product because the majority of the fumonisin is located in the pericarp. At present, there is a need for further studies that combine the monitoring of fumonisin content during grain processing with the determination of the efficiency of pericarp removal.

Chemical Changes

Alkaline cooking of corn is likely to produce strong interactions between calcium ions and starch. At alkaline pH, partial ionization of hydroxyl groups on starch can occur, enhancing interaction with calcium ions (*4*). The amount of calcium remaining in the processed corn is dependent on the concentration of lime used for processing and the length of time spent cooking and steeping (*5*)

Corn tortillas are nutritionally desirable because they have low fat, high calcium and high levels of dietary fiber. Niacin is also more bioavailable as a result of the nixtamalization. In an industrial process for producing soft tortillas, the moisture content increases progressively from corn (12%), as nixtamal (48%) and masa (53%) are produced (Figure 1). During baking, approximately 10-12% moisture is lost from the masa, resulting in tortillas of 38-46% moisture.

As a result of nixtamalization, if mycotoxin fumonisin B_1 (FB_1) is present, it will undergo hydrolysis, with the removal of the 1,2,3-propanetricarboxylic acid sidechains at C-14 and C-15 to form hydrolyzed FB_1 (HFB_1), shown in Figure 2. An earlier report had suggested that HFB_1 had a toxicity equal to or greater than that of FB_1 (*6*), but recent research has indicated that HFB_1 is at least five-fold less toxic than FB_1 (*7, 8*), as determined by inhibition of the enzyme ceramide synthase. There is also the possibility that fumonisin will react with glucose during nixtamalization in a manner analogous to that describing the formation of *N*-(carboxymethyl) fumonisin B_1 under alkaline conditions at 78 °C (*9*).

Fumonisins

The fungal pathogens, *Fusarium verticilliodes* (Sacc.) Niremberg (synonym *F. moniliforme* Sheldon) and *F. proliferatum*, are found in corn worldwide, occurring naturally within the corn plant, but these fungi can also be introduced by insect damage (*10*). Under certain circumstances, these fungi produce fumonisins. Even apparently healthy grains of corn can contain moderate amounts of fumonisin. There is concern that a high dietary intake of corn-based foods may be exposing certain human populations to fumonisins.

A variety of factors will impact the degree to which human exposure to fumonisin will occur: (1) high levels of fumonisins have been associated with hot and dry weather that is followed by periods of high humidity; (2) storage conditions can promote increased levels of fumonisins if the moisture content of the harvested corn is optimal for growth of fumonisin-producing fungi; (3) diets containing high amounts of corn will increase the potential for exposure to fumonisin-contaminated food; and (4) the amount of fumonisin present may be reduced by the methods involved in food processing. For example, higher levels of fumonisin are detected in corn screenings, broken kernels which are screened from bulk corn prior to processing. Removal of broken pieces by cleaning (screening) of corn can reduce the amount of fumonisin by 30-40 %, compared to the amount detected in unscreened corn. Corn screenings are often used in animal feeds, not for tortilla production.

209

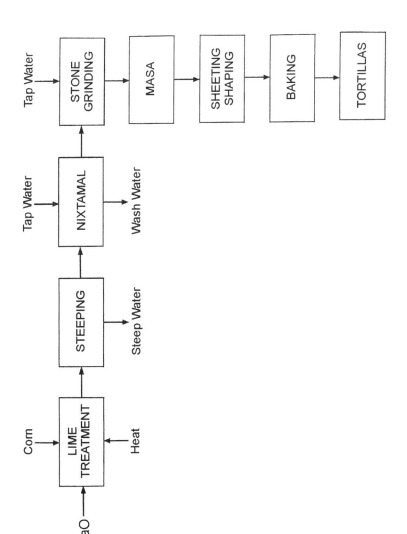

Figure 1. Schematic illustration of the nixtamalization process: production of tortillas using alkaline hydrolysis.

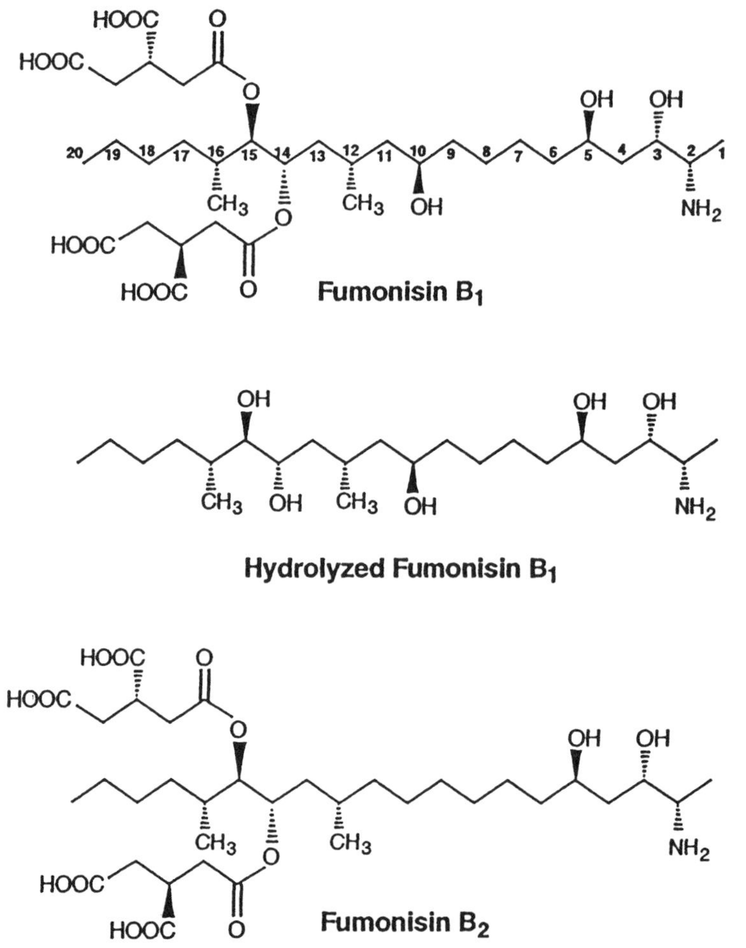

Figure 2. Absolute configuration of fumonisin B₁, hydrolyzed fumonisin B₁ and fumonisin B₂.

Fumonisins represent a family of structurally related compounds. Fumonisin B_1, shown in Figure 2, is the most prevalent of the fumonisins in naturally contaminated corn and is usually present as 70% of the total fumonisins detected. Fumonisins inhibit sphingolipid biosynthesis by interfering with the enzyme sphinganine (sphingosine) N-acyltransferase (ceramide synthase), resulting in accumulation of sphinganine and decreased biosynthesis of ceramides and complex sphingolipids (11). Elevated levels of sphinganine occur after both *in vivo* and *in vitro* exposure to fumonisins (12).

Because FB_1 can occur in corn and corn-based products in the United States and has been associated with toxic effects in horses and pigs (13, 14), the National Toxicology Program (NTP) undertook a two-year study in which male and female F344/N rats and B6C3F$_1$ mice were exposed to FB_1 for two years. A recent release of the Draft of the Technical Report (NTP TR 496) (15) showed clear evidence of carcinogenic activity of FB_1 in male F344/N rats as indicated by increased incidences of renal tubule neoplasms, as well as clear evidence of carcinogenic activity of FB_1 in female B6C3F$_1$ mice based on the increased incidences of hepatocellular neoplasms. It should be noted that the amount of FB_1 ($\geq$ 50 ppm) in the diets of the rats and mice was much higher than the amount detected in corn in the United States. The final version of NTP Technical Report 496 is currently in press.

Current data on the occurrence of fumonisins in corn-based human food has indicated that the levels are quite low. The corn dry milling industry recently completed a three-year voluntary monitoring program for fumonisin content (FB_1 + FB_2 + FB_3) in white and yellow corn meal. Examination of 1,562 samples indicated that greater that 95% of the samples had less than 1 ppm fumonisin (16).

In a recent study, we evaluated the ability of nixtamalization to reduce the amount of fumonisin in naturally contaminated corn (Table I) (17). Previous studies of masa and tortillas (18) produced in Mexico had indicated that the amount of FB_1 present was significantly higher, compared to samples purchased in the United States (means of 0.79 ppm and 0.16 ppm, respectively). Because the concentration of fumonisins in the unprocessed maize was not known, it was not possible to determine whether higher levels of fumonisin were present in the Mexican maize before processing or if the maize had been incompletely nixtamalized.

In the pilot scale nixtamalization (Figure 1) of fumonisin-contaminated corn (FB_1, 8.8 ppm; FB_2, 2.0 ppm), Grade 2 yellow dent corn, free of cracked kernels and without significant mold or insect damage, was used for processing (17). The kernel characteristics (test weight, 59 lb/bu; density, 1.3 g/cm^3) of the corn processed represented the traits of an ideal food corn (19). A special effort was made to use normal-appearing corn as would be used in commercial operations. The initial weight of each component, corn moisture and wastewater solids were measured at each stage for mass balance calculation.

**Table I. Fumonisin Content in Solid Fractions Produced by
Nixtamalization of Fumonisin-Contaminated Corn**

Sample	FB_1	HFB_1	FB_2
Corn	8790 (22.1)[a]	ND	1970 (21.3)
Rep I			
Steeped nixtamal	111	229	21
Washed nixtamal	172	197	75
Masa	322	304	50
Tortillas	406	282	100
Rep II			
Steeped nixtamal	867 (15.6)	576 (28.6)	555 (1.1)
Washed nixtamal	1075 (26.8)	579 (20.4)	708 (42.6)
Masa	500 (33.4)	366 (19.3)	193 (1.9)
Tortillas	602 (22.1)	445 (21.2)	315 (28.5)

[a] Results are expressed as ng/g. Data for Rep I are single determinations. Data for corn and Rep II are means (n=2 for corn, steeped nixtamal and masa; n=4 for washed nixtamal and tortillas) (%RSD). Samples were dried, ground and extracted; fumonisins were detected by HPLC, following derivatization with naphthalene dicarboxaldehyde (NDA). Adapted from Reference 17. Copyright 2000 American Chemical Society.

Development of Methods for Extraction and Detection of Fumonisins

Different methods have been developed to optimize the extraction of fumonisin from nixtamalized corn products. In addition, methods have been developed to address the simultaneous detection of both FB_1 and HFB_1. It is not possible to use methods (strong anion exchange clean-up columns) that have been developed to measure unhydrolyzed fumonisins in corn for quantitation of hydrolyzed fumonisins because the former will adhere to the column, but the latter will not. It is important to state that fumonisins are suspected carcinogens and should be handled with care.

Means of detection for the fumonisins include HPLC, MS and ELISA. Because fumonisins do not have any intrinsic absorbance or fluorescence, they are derivatized with fluorescent probes for detection by HPLC. A primary amine group is required for derivatization of fumonisin. Additionally, specific antibodies capable of recognizing either the parent fumonisin (FB_1) or the hydrolyzed form (HFB_1) have been developed for use in affinity chromatography and ELISA (20).

Different solutions have been used to optimize efficient extraction of fumonisins from nixtamalized products. Use of EDTA in the extraction solvent, acetonitrile-0.01 M EDTA (1:1), has proven to be advantageous for extracting fumonisins from alkali-treated corn (*21*). In these studies, alkali treatment of fumonisin-contaminated corn was performed at room temperature, as opposed to commercial processing where corn is boiled in lime and allowed to steep overnight. Those kernels, from which pericarps were fully removed, only retained 5.1 % of the original FB_1. When the calcium content of the processing samples was measured and the amount of EDTA was adjusted so that it exceeded by 1.36 on a molar basis the calcium concentration, 25 % additional FB_1 was extracted (*18*).

Methods for extraction of fumonisins have also included mixtures of organic solvents and use of elevated temperatures or acid conditions. Much better recovery of FB_1 was obtained when methanol-acetonitrile-water (1:1:2), instead of methanol-water (8:2), was used to extract FB_1 from naturally contaminated corn that had been alkali processed (*22*). Results using the methanol-acetonitrile-water solvent mixture were similar to those obtained with acetonitrile-water containing EDTA in molar excess (*18*). The quantity of fumonisins extracted from nacho chips and taco shells increased at elevated temperature (80°C), using solvent mixtures of methanol-acetonitrile-water (1:1:2) or ethanol-water (8:2) (*23*). Even water at elevated temperatures appears to be an effective extraction solvent. It is anticipated that similar efficiency in extraction of fumonisins would occur at 80°C using acetonitrile-water (1:1). Use of acidic conditions also increased the amount of fumonisin extracted; the total fumonisin content of two tortilla products increased 95 % and 495 %, respectively, when methanol-0.1 M hydrochloric acid (3:1) was used instead of methanol-water (3:1) (*24*).

Fumonisins in Tortillas and Related Products

A number of reports have described the amount of fumonisin present in nixtamalized corn products. Results indicate that FB_1 can frequently be detected in corn-containing foods. It is important to keep in mind that samples of corn from different years can have different levels of fumonisin present, with exposure to environmental factors (drought stress) being responsible for higher fumonisin levels. Additionally, because only the finished products have been analyzed, it is not possible to predict how efficient processing has been in reducing the amount of fumonisin that had been present in the starting material.

Analyses of tortillas and masa from the Texas-Mexico border indicated that both FB_1 and its hydrolysis product were present. In tortillas, the average amounts of FB_1 and HFB_1 were 0.187 ppm and 0.082 ppm, respectively. Average amounts in masa were 0.262 ppm and 0.064 ppm for FB_1 and HFB_1, respectively (*25*). The amount of FB_1 was found to be significantly higher in Mexican samples of masa and tortillas (mean of 0.79 ppm) than in samples purchased in the United States (mean of 0.16 ppm) (*18*).

Examination of samples purchased in markets in Germany indicated that nearly all alkali-processed corn food products contained both FB_1 and HFB_1, but the amounts of FB_1 (0.039-0.185 ppm) were usually higher than the level of HFB_1 (0.044-0.083 ppm) present (*26*). In samples of nixtamalized corn (tortillas and nixtamal) from Guatemala, both FB_1 and HFB_1 were detected; the highest amounts of FB_1 and HFB_1 present were 11.6 ppm and 185 ppm, respectively (*27*). It was not possible to determine the degree to which fumonisin content was reduced by nixtamalization because there were no data linking samples from different stages of the processing. The levels of fumonisin detected in the nixtamalized corn suggest that rather high levels were present in the raw corn from Guatemala.

Effects of Processing on Reduction of Fumonisins

Processing can reduce the level of fumonisins, make it more difficult to detect fumonisins or have no effect. Fumonisins are stable at ambient temperature in a pH range of 4-10, as well as at temperatures up to 100 °C at neutral pH (*28*). It is when fumonisins are exposed to the combination of alkaline pH and elevated temperature (100 °C) that fumonisins are hydrolyzed, leeching fumonisins into the liquid fractions (Table II) (*17*). When corn naturally contaminated with fumonisin was processed by nixtamalization to determine how much of the original fumonisin would be present in the finished products (masa and tortillas), the amount of fumonisin was reduced by > 81% (Table I) (*17*).

Table II. Fumonisin Content in Aqueous Fractions Produced by Nixtamalization of Fumonisin-Contaminated Corn

Sample	FB_1	%	HFB_1	FB_1 equiv	%
Steep water					
Rep I	267 (15.4)[a]	3.0[b]	3041 (3.1)	5414	61.6
Rep II	ND[c]	ND	3753 (18.2)	6681	76.0
Wash water					
Rep I	360 (4.0)	4.1	153 (6.7)	272	3.1
Rep II	ND	ND	211 (6.4)	375	4.3

[a] Results for FB_1, HFB_1 and FB_1 equiv are expressed as ng/g. Data are means (n=3) (%RSD). [b] % of original FB_1 present in corn. [c] ND, not detected (<10 ng/g). Adapted from Reference 17. Copyright 2000 American Chemical Society.

Recently, a preliminary report on the effect of a commercial tortilla process on fumonisin content showed that the amount of fumonisin was reduced by 40-80 %, following cooking, soaking and washing (29), in agreement with the data shown in Tables I and II (17). More calcium was retained in Rep II fractions; this may have been responsible for higher levels of fumonisin detected (Table I). Sheeting, baking and frying at commercial times and temperatures did not reduce fumonisin further (29), in accord with the results for tortillas (Table I). Tortillas contained approximately 0.50 ppm FB_1, plus 0.36 ppm HFB_1. This represented 18.5 % of the initial FB_1 concentration (17). Similar reduction of aflatoxin (> 83 %) occurred after corn had undergone nixtamalization (30). On the other hand, traditional fermentation for producing Nepalese maize beer did not eliminate fumonisins, although it was possible by hand-sorting visibly diseased kernels to detoxify contaminated maize (31).

Summary

Compared to raw corn, the amount of FB_1 and FB_2 present in tortillas, the finished product, represented substantial reductions. Most of the loss occurred in the process of steeping and washing, with 76 % of the fumonisin content remaining in the liquid fractions. Thus, the traditional alkaline cooking technique, nixtamalization, appears to be a means of significantly reducing the amount of fumonisin found in corn.

As alternate means of reducing fumonisin in corn, sources of fumonisin detoxifying enzymes have been sought. *Exophiala spinifera*, a black yeastlike fungus, has been shown to transform FB_1 through the activity of a soluble extracellular esterase to the amino polyol AP_1 (analogous to HFB_1), which can undergo oxidative deamination (32). The enzymatic removal of the amine group of fumonisin is an important means of detoxification because a free amine is thought to be critical for biological activity of FB_1 or AP_1

There is no conclusive information, at the present, linking adverse human health effects with fumonisins. Although human epidemiological studies are presently inconclusive, the Food and Drug Administration (FDA) believes that there is the possibility of human health risks because a variety of significant adverse health effects associated with fumonisins have occurred in livestock and experimental animals. Consequently, FDA has distributed a draft guidance document for comment purposes only regarding the maximum recommended levels of fumonisins for corn used in production of human foods and animal feeds (33). The recommended level for total fumonisins (FB_1 + FB_2 + FB_3) for cleaned corn intended for masa production has been set at 4 ppm. It is believed that typical fumonisin levels in corn and corn products destined for human consumption are actually much lower than the recommended levels.

216

References

1. Rooney, L. W.; Serna-Saldivar, S. O. In *Corn: Chemistry and Technology*, Watson, S. A., Ramstad, P. E., Eds.; American Association of Cereal Chemists, Inc. St. Paul, MN, 1987; pp 399-429.
2. Campus-Baypoli, O. M.; Rosas-Burgos, E. C.; Torres-Chavez, P. I.; Ramirez-Wong, B.; Serna-Saldivar, S. O. *Starch* **1999**, *51*, 173-177.
3. Rooney, L. W.; Suhendro, E. L. *Cereal Foods World* **1999**, *44*, 466-470.
4. Bryant, C. M.; Hamaker, B. R. *Cereal Chem.* **1997**, *74*, 171-175.
5. Serna-Saldivar, S. O.; Rooney, L. W.; Greene, L. W. *Cereal Chem.* **1991**, *68*, 565-570.
6. Hendrich, S.; Miller, K. A.; Wilton, T. M.; Murphy, P. A. *J. Agric. Food Chem.* **1993**, *41*, 1649-1654.
7. Schmelz, E. M.; Dombrink-Kurtzman, M. A.; Roberts, P. C.; Kozutsumi, Y.; Kawasaki, T.; Merrill, A. H., Jr. *Toxicol. Appl. Pharmacol.* **1998**, *148*, 252-260.
8. Norred, W. P.; Plattner, R. D.; Dombrink-Kurtzman, M. A.; Meredith, F. I.; Riley, R. T. *Toxicol. Appl. Pharmacol.* **1997**, *147*, 63-70.
9. Howard, P. C.; Churchwell, M. I.; Couch, L. H.; Marques, M. M.; Doerge, D. R. *J. Agric. Food Chem.* **1998**, *46*, 3546-3557.
10. Nelson, P. E.; Desjardins, A. E.; Plattner, R. D. *Annu. Rev. Phytopathol.* **1993**, *31*, 233-252.
11. Wang, E.; Norred, W. P.; Bacon, C. W.; Riley, R. T.; Merrill, A. H., Jr. *J. Biol. Chem.* **1991**, *266*, 14486-14490.
12. Merrill, A. H., Jr.; Liotta, D. C.; Riley, R. T. *Trends Cell Biol.* **1996**, *6*, 218-233.
13. Marasas, W. F. O.; Kellerman, T. S.; Gelderblom, W. C. A.; Coetzer, J. A. W.; Thiel, P. G.; van der Lugt, J. J. *Onderstepoort J. Vet. Res.* **1988**, *55*, 197-203.
14. Harrison, L. R.; Colvin, B. M.; Greene, J. T.; Newman, L. E.; Cole, R. J. *J. Vet. Diagn. Invest.* **1990**, *2*, 217-221.
15. National Toxicology Program, 1999, Draft Technical Bulletin 496, URL http://ntp-server.niehs.nih.gov/htdocs/lt-studies/tr496.html
16. Anon. *Milling J.* 1999, July-September 1999, pp 36-37.
17. Dombrink-Kurtzman, M. A.; Dvorak, T. J.; Barron, M. E.; Rooney, L. W. *J. Agric. Food Chem.* **2000**, *48*, 5781-5786.
18. Dombrink-Kurtzman, M. A.; Dvorak, T. J. *J. Agric. Food Chem.* **1999**, *47*, 622-627.
19. Serna-Saldivar, S. O.; Gomez, M. H.; Rooney, L. W. In *Advances in Cereal Science and Technology*; Pomeranz, Y., Ed.; American Association of Cereal Chemists, Inc. St. Paul, MN, 1990; Vol. 10, pp 243-307.

20. Maragos, C. M.; Bennett, G. A.; Richard, J. L. *Food Agric. Immunol.* **1997,** *9,* 3-12.
21. Sydenham, E. W.; Stockenstrom, S.; Thiel, P. G.; Shephard, G. S.; Koch, K. R.; Marasas, W. F. O. *J. Agric. Food Chem.* **1995,** *43,* 1198-1201.
22. Scott, P. M.; Lawrence, G. A. *Food Addit. Contam.* **1996,** *13,* 823-832.
23. Lawrence, J. F.; Niedzwiadek, B.; Scott, P. M. *J. AOAC Int.* **2000,** *83,* 604-611.
24. Meister, U. *Mycotoxin Res.* **1999,** *15,* 13-23.
25. Stack, M. E. *J. AOAC Int.* **1998,** *81,* 737-740.
26. Hartl, M.; Humpf, H. U. *J. Agric. Food Chem.* **1999,** *47,* 5078-5083.
27. Meredith, F. I.; Torres, O. R.; Saenz de Tejada, S.; Riley, R. T.; Merrill, A. H., Jr. *J. Food Prot.* **1999,** *62,* 1218-1222.
28. Jackson, L. S.; Hlywka, J. J.; Senthil, K. R.; Bullerman, L. B.; Musser, S. M. *J. Agric. Food Chem.* **1996,** *44,* 906-912.
29. Saunders, S. Fumonisins Risk Assessment Workshop (Abstracts) 2000, P 35.
30. Guzman de Pena, D.; Trudel, L.; Wogan, G. N. *Bull. Environ. Contam. Toxicol.* **1995,** *55,* 858-864.
31. Desjardins, A. E.; Manandhar, G.; Plattner, R. D.; Maragos, C. M.; Shrestha, K.; McCormick, S. P. *J. Agric. Food Chem.* **2000,** *48,* 1377-1383.
32. Blackwell, B. A.; Gilliam, J. T.; Savard, M. E.; Miller, J. D.; Duvick, J. P. *Nat. Toxins* **1999,** *7,* 31-38.
33. U. S. Food and Drug Administration, Center for Food Safety and Applied Nutrition, Center for Veterinary Medicine, 2000, Draft Document (Guidance for Industry), URL http://vm.cfsan.fda.gov/~dms/fumongui.html

Chapter 16

Potassium Bromate in Bakery Products: Food Technology, Toxicological Concerns, and Analytical Methodology

Gregory W. Diachenko and Charles R. Warner

Division of Product Manufacture and Use, U.S. Food and Drug
Administration, Washington, DC 20204

Potassium bromate, which has been used since 1916 as a
dough conditioner, has been found, in some circumstances, to
leave bromate residues in retail bakery products. Studies
published in 1983 established potassium bromate as an
animal carcinogen. Over the past decade, improvements in
analytical methodology have made it possible to detect these
residues at levels of 5 µg/kg (ppb). Research within
laboratories of the baking industry has established that
reducing the amount of added potassium bromate,
maintaining higher baking temperatures, and using ferrous
salts as the nutritive iron supplement will significantly reduce
or eliminate bromate residues. FDA surveys have revealed
progress towards the goal of eliminating residues; however,
more work is required.

This paper will review the use of potassium bromate in baked goods with
emphasis on the regulatory responsibilities of the Food and Drug
Administration (FDA), the toxicological concerns regarding potassium

bromate, the recent developments in analytical methodology that have revealed the presence of bromate residues in finished baked products, and the research program carried out by the baking industry on baking practices that will minimize bromate residues. The results of a mini survey of baked products from commercial outlets within the Washington DC area at two different times will be presented to gauge the progress by the industry. In addition, preliminary exposure data and risk estimates are presented. Future activities by both the baking industry and the Food and Drug Administration will have a profound effect on whether or not potassium bromate will continue to be permitted in baking operations.

The Need for Regulatory Action

Potassium bromate and calcium bromate have been used as dough conditioners since 1916 (1). They are additives identified in FDA's food standards for bread, dough, and flour and can be added to the flour or dough in amounts up to 75 ppm (2). In 1982 and 1983 Kurokawa et al. (3,4) published the results of animal tests that demonstrated that potassium bromate, administered in the drinking water, causes renal cancer in rats. More recently Wolf et al. (5) have reported that potassium bromate is a rodent carcinogen and a nephro- and neurotoxicant in humans. However, it was the combination of the initial report by Kurokawa (3) and the development of analytical methods that revealed the presence of residues of bromate in finished baked goods that stimulated regulatory concern in many countries. In 1989, the United Kingdom (UK) banned potassium bromate uses as a dough conditioner in bakery products (6).

As a result of the demonstrated carcinogenicity, the findings of bromate residues in retail bakery goods and the regulatory action by the UK, the FDA met with the baking industry in 1991 and requested that they begin phasing out the use of potassium bromate. The industry has studied the optimization of baking parameters to reduce bromate residues (7) and investigated a number of alternative dough conditioners such as potassium iodate, ascorbic acid and azodicarbonamide (8). In addition, analytical methods that were needed to provide the data to accurately assess the prevalence of bromate residues and to monitor the progress of food technological studies, have been developed in industry laboratories. These efforts have yielded data which the industry hopes will convince the FDA that there is no need to ban the use of potassium bromate as was done in the UK and Canada (6,9). Analytical chemistry has played a very important role in the FDA's deliberations.

Analytical Methodology

The evolution of the analytical methodology over the past decade has been very interesting. The methods that initially revealed the presence of bromate residues were based upon the bromate oxidation of bromide to bromine that added across the double bond in styrene to yield a bromoalkyl side chain on benzene that was analyzed by gas chromatography with electron capture detection (*10,11*). While these methods provided valuable information in the early stages of investigations on potassium bromate residues, an HPLC method that was developed by industry has proven to be the current method-of-choice (*12*). This method, which was published in 1997 utilizes an ion pairing reverse phase chromatographic method with a post-column derivatization, can be used to detect residues as low as 5 ng/g. This is far lower than previous methodology (*10*) that was used in the UK to reveal the presence of potassium bromate residues in bakery goods back in the late 80's. Himata et al. (*13*) successfully completed a rigorous method trial with the FDA laboratory as the peer laboratory. Blind samples with potassium bromate fortifications ranging from 10 ppb to 52 ppb in white bread; multigrain bread and coffee cake were submitted to the peer laboratory. The results were satisfactory and the AOAC International has awarded the method official Peer-Verified$^{©}$ status (*13*).

To achieve a limit of detection (LOD) of ca. 5 ppb the procedure incorporates multiple steps that begin with extracting 10 grams of bread with 50 ml of water in a blender system. This aqueous extract is passed through a 500 mg C-18 solid phase extractor (SPE) to remove lipophilic co-extractives. The chloride, which is typically present in the aqueous extract at concentrations approximately five to six orders of magnitude greater than the concentration of bromate, is removed by passing the extract through a SCX column with an ion exchange capacity of 2.0-2.5 meq/cartridge in the silver ion form. The silver chloride precipitate is removed by filtration. The aqueous extract, which at this stage is essentially water-clear, is subjected to ultrafiltration to remove proteins and polysaccharide extractives with molecular weights in excess of 10K daltons. Some silver ion, which will cause deterioration of the HPLC column, inevitably leaches into the aqueous solution; therefore, further treatment of the aqueous extract with a SCX column in the sodium form is required to remove these traces of soluble silver ions. The extract is ready for HPLC analysis with a 4.6 mm x 250 mm Zorbax SB-C18 column with mobile phase consisting of 10% methanol in an aqueous solution of 0.05 M tetrabutylammonium acetate at ca. pH6. As previously described for water analysis (*14*) the bromate in the eluent is detected spectrometrically at 450 nm after a post-column reaction with acidified *o*-dianisidine. A flow chart of this methodology is shown in Figure 1.

The method recoveries of bromate, that was added to unbromated bread, range from about 73 ±11.5% at 5 ng/g to 86.5±5.1% at 100 ng/g (*13*). The overall mean recovery was 83%±7.6% with a relative standard deviation of 9.6%. This methodology was used by the baking industry to evaluate processing

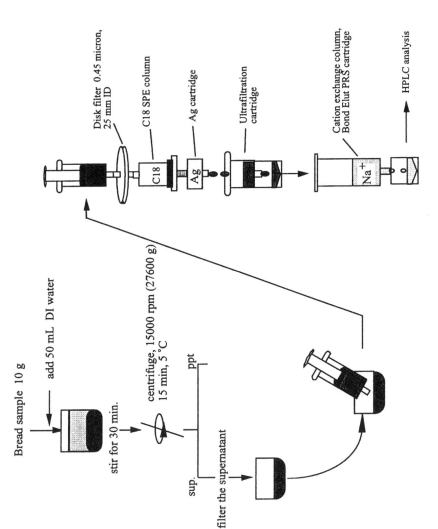

Bread sample 10 g

add 50 mL DI water

stir for 30 min.

centrifuge, 15000 rpm (27600 g)
15 min, 5 °C

ppt

sup.

filter the supernatant

Disk filter 0.45 micron,
25 mm ID

C18 SPE column

Ag cartridge

Ultrafiltration
cartridge

Cation exchange column,
Bond Elut PRS cartridge

HPLC analysis

Figure 1. Analytical Flow-Chart. (reproduced with permission from reference 12).

changes that might be implemented to control or, preferably, eliminate bromate residues in bakery products.

Baking Practices that Reduce or Eliminate Bromate Residues

Yamazaki Bakery of Tokyo, Japan and other members of the baking industry have studied readily identifiable parameters that could be controlled to eliminate or at least reduce potassium bromate residues in the final baked products to non-detectable levels, i.e., less than 3 to 6 ppb (7). Thewlis noted that dough seems to have a finite level of reductants and potassium bromate added above 75 ppm leads to very high residues (15). It was found that reducing the added potassium bromate to 25 ppm for loaf bread resulted in elimination of detectable bromate residues (13). To have the same success with rolls and buns it was necessary to reduce the amount of added potassium bromate to 15 ppm because, in contrast to loaves of bread, the internal temperatures of rolls and buns do not reach full oven temperatures. Industry studies have shown that higher baking temperatures effectively reduce bromate residues. The key manufacturing parameters are summarized in Table I (7).

Table I. Processing Changes to Control Bromate Residues in Bakery Products

- Add <30 ppm PB for breads
- Add <15 ppm for rolls and buns
- Use longer baking times and higher temperatures
- Add 100 ppm ascorbic acid
- Use bleached flour
- Use ferrous sulfate
- Avoid adding azodicarbonamide

Industrial research has demonstrated that judicious incorporation of other ingredients can also reduce bromate residues (7). For example, it was discovered that the use of bleached flour instead of unbleached flour was associated with lower bromate residues. Not surprisingly, utilizing ascorbic acid at 100 ppm as a dough conditioner will reduce bromate residues. The selection of ferrous sulfate as the iron nutritive supplement, rather than elemental iron, contributes to the goal of reducing bromate residues. Clearly, the industry has made great inroads in understanding the key parameters that influence bromate residues and is working to introduce these baking practices as general guidelines for the industry. To obtain an indication of the effectiveness of the industry in reducing or eliminating residues, in 1998-1999 the FDA analyzed for bromate in baked goods purchased in the Washington, DC area.

Market Surveys

The bakery products were collected at retail outlets and the results are given in Table II.

Table II. 1998-1999 Survey of Potassium Bromate Levels in Baked Goods

Brands	Description	$KBrO_3$ (ppb)
Brand A	Hamburger Buns	240*
	Hamburger Buns	39
Brand B	Deli Steak Rolls	42*
	Hamburger Buns	48
	Hamburger Buns	ND
	Dinner Rolls	19
Brand C	Hot Dogs	ND*
Brand D	Hamburger Buns	290*
Brand E	Brown & Serve Buns	27
	Hamburger Buns	55
Brand F	Enriched Hot Dog Rolls	40
Brand G	Grinder Rolls	<5
	French Club Rolls	5
	French Club Rolls	ND
Brand H	Hoagie Rolls	2840
Brand I	Kaiser Rolls	ND
Brand J	Rolls	1144

*Center cores that did not include the crusts were taken for analysis.

Primarily, rolls and buns were selected because these typically have the highest levels of potassium bromate. This strategy, which admittedly yields results biased on the high side, was adopted because it was felt that it would give the best snap shot of bromate residues in baked goods. The greatest challenge for the industry is to eliminate residues in these single-serving products.

Table II reveals that there is a wide range of bromate levels detected in this survey that took place over the 1998-1999 calendar years using the HPLC method. (13). The levels range from non-detectable at less than ca. 5 ppb all the way up to the highest product with 2840 ppb in brand H hoagie rolls. These data were compared to the 1992-93 survey that included retail outlets in the

Washington DC area as well as a number of surrounding areas. Again, the data is too limited to draw more than inferences from the results. However, the results shown in Table III, as well as Figure 2, suggest that there has been some reduction in bromate residues over the last decade.

Table III. Comparison of Bromate Residues (ppb) in Products Surveyed in 1998-1999 vs 1992-1993

	1998-1999 Survey	*1992-1993 Survey*
Average KBrO3	282	485
Median	39	68
Standard Deviation	715	1208
Number of Products Tested	17	25

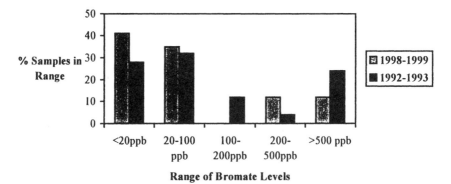

Figure 2. Comparison of Bromate Residues in Products Surveyed in 1998-1999 vs 1992-1993

In the more recent survey in 98-99 there is a definitive increase in the percentages of samples that were in the less than the 20 ppb range. The significance of 20 ppb is discussed later in the risk assessment section. The products falling in the less than 20 ppb category went from 28% in the 92-93 survey up to about 42% in the 98-99 survey. The 92-93 survey had many more samples in the 100-200 ppb range. The 98-99 survey found a much higher percentage in the 200-500 ppb range, but that was due to one manufacturer who has already been the subject of regulatory action because of the use of excessively high levels of potassium bromate in the dough. The statistical summary of the data given in Table III does suggest that the industry is having some success in reducing bromate residues. The results in Table III for the 98-99 survey and the 92-93 survey indicate an average of 282 ppb (median of 39 ppb) and an average of 485 ppb (median of 68 ppb), respectively. The very large standard deviation for the 92-93 survey is primarily due to a few very high level outliers. The difference between surveys is too limited to pass the test of statistical significance because of the relatively small number of products; 17 in 98-99 and 25 in 92-93. Manufacturer H, who supplied the outlier in the 98-99 survey results, has actually stopped using potassium bromate. Therefore, eliminating that high value from the results for the 98-99 survey would lower the average potassium bromate level almost 10 fold. This provides a more encouraging picture of the effectiveness of the industry's efforts to reduce bromate levels.

Risk Assessment and Exposure Estimates

In order to put these residue levels in perspective a quantitative risk assessment was carried out by the Quantitative Risk Assessment Committee (QRAC) of the Center for Food Safety and Applied Nutrition. They used a mean intake of baked goods of 56 grams per person per day (g/p/d) and a 90% intake for the heavy consumer of baked goods of 105 grams per person per day (g/p/d). These numbers came from the 1989-91 USDA Continuing Survey of Food Intakes by Individuals. Using the 1992-93 survey data on bromate levels given in Figure 2 and Table III, we can estimate a mean exposure to bromate of about 10 µg/p/d from baked goods. Multiplying the QRAC calculated unit risk factor of 3.0×10^{-2} (mg/kg-body weight (bw)/d)$^{-1}$ (16) times a mean exposure of 0.17 µg/kg-bw/d (obtained by dividing the 10/ug/p/d exposure by an average body weight of 60 kg per person); we obtain a lifetime upper-bound cancer risk estimate for average exposure to bromate from baked goods based on the 92-93 survey data of about 5×10^{-6}. If you use the 98-99 survey data there would be a proportionally lower risk since you are at about half the bromate exposure level. What also came out of this preliminary risk assessment was the estimation of 20 ppb bromate as the one in a million upper bound cancer risk level for consumers at the 90th percentile intake of baking products, which is often used as an insignificant risk level. This 20 ppb level has been used as an informal target for the industry and those who are using potassium bromate have been encouraged to take every step necessary to keep their final product residues well under 20 ppb.

The Future of Potassium Bromate as a Dough Conditioner

As far as future activity is concerned, the baking industry has informed the agency that they are continuing to refine the baking parameters that will eliminate bromate residues and to distribute this information as good manufacturing practices guidance to those bakeries that choose to continue to use potassium bromate. An active monitoring program must be established and corrective action taken quickly if they find products that exceed 20 ppb. Based upon experience in the past, particular attention must be paid to buns and rolls. Some bakers will decide to manufacture baked goods without potassium bromate, but many companies, in the belief that residues can be reduced or eliminated, are going back to potassium bromate. Therefore, there is a need for FDA to continue to monitor industry progress in controlling bromate levels and evaluate its regulatory and voluntary compliance options based on the ability of the industry to reliably eliminate any significant risk to the consumer.

References

1. Fitchett, C.S.; Frazier, P. In *Chemistry and Physics of Baking: Materials, Processes, and Products*; Blanshard, J. M. V.; Frazier, P. J.; Galliard, T., Eds.; Royal Society of Chemistry: London, 1986; Chap. 14, pp 179-198.
2. 21 Code of Federal Regulations; §§ 136.110(c)(14)(i), 137.155, 137.160, 137.205.
3. Kurokawa,Y.; Hayashi, Y.; Maekawa, A; Takahashi, M.; Kokubo, T. *Gann.* **1982**, *73*, 335-338.
4. Kurokawa,Y.; Hayashi, Y.; Maekawa, A.; Takahashi, M.; Kokubo, T., Odashima, S. *J. Nat'l Cancer Inst.*. **1983**, *71*, 965-972.
5. Wolf, D. C.; Crosby, L. M.; George, M. H.; Kilburn, S. R.; Moore, T. M.; Miller, R. T.; DeAngelo, A. B. *Toxicol. Pathol.* **1998**, *26*, 724-729.
6. Clare, C. *BIBRA Bulletin*, **1989**, 474-476
7. Giesecke, A.G.; Taillie, S.A. *Cereal Foods World*, **March 2000**, *45*, 111-120.
8. Ranum, P. *Cereal Foods World*, **March 1992**, *37*, 253-258
9. Jackel, S.S. *Baking*, **1994**, 39, 772
10. Dennis, M. J.; Burrell, A.; Mathieson, K.; Willetts, P.; Massey, R. C. *Food Additives and. Contaminants.* **1994**, *11*, 633-639.
11. Himata, K.; Kuwahara, T.; Ando, S.; Maruoka, H. *Food Additives and. Contaminants.* **1994**, *11*, 559-569.
12. Himata, K.; Noda, M.; Ando, S.; Yamada, Y. *Food Additives and. Contaminants.* **1997**, *14*, 809-818.

13. Himata, K.; Noda, M.; Ando, S.; Yamada, Y. *J. AOAC International.* **2000**, *83*, 347-355.
14. Warner, C. R.; Daniels, D. H.; Joe, F. L.; Diachenko, G. W. *Food Additives and Contaminants.* **1996**, *13*, 633-638.
15. Thewlis, B. H. *J. Sci. Food Agric.* **1974**, *25*, 1471-1475.
16. "Potassium Bromate"; Internal FDA CFSAN Memorandum dated May 10, 1988, from Quantitative Risk Assessment Committee, to W. Gary Flamm, Director, Office of Toxicological Sciences, Center for Food Safety and Applied Nutrition, U. S. Food and Drug Administration, Washington, DC 20204, 1988.

Chapter 17

Effect of Food Processing on Bioactive Compounds in Foods: A New Method for Separation and Identification of *cis*-Cinnamic Acid from Its Racemic Mixture

Fang-Ming Sun[1], James L. Smith[2], B. M. Vittimberga[2], and R. W. Traxler[3]

[1]Department of Health and Nutrition, Chia Nan College of Pharmacy and Science, Tainan 71710, Taiwan
[2]Department of Chemistry, University of Rhode Island, Kingston, RI 02881
[3]Department of Biochemistry, Microbiology, and Molecular Genetics, University of Rhode Island, Kingston, RI 02892

A HPLC method was developed to separate racemic mixtures of cinnamic acid in this study. The cis-isomer obtained by this method was confirmed and characterized by using GC-mass, FT-IR, and one dimensional proton NMR. It was found the resulting data from mass pattern, FT-IR spectra, and J (coupling constant) for the differences between cis- and trans-cinnamic acid can be used as an additional and important index for distinguishing these two racemic mixtures. The stability of the cis-isomer prepared by this study has been shown to be more than one year suggesting that the presenting HPLC preparation is a better and more precise method than the previous method of using ion exchange chromatography.

Introduction

Cinnamic acid is present in all kinds of plant derived foods, herbs, and medicines (Wolffram et al. 1994, Chen and Sheu 1995). Two forms of cinnamic acids, trans-cinnamic acid and cis-cinnamic acid, have been found to exist in the plant cells ($_{van}$ Overbeek 1951). However, the trans-cinnamic acid has been shown to be the predominate form in nature (>99%), since it is much more stable than the cis-isomer ($_{van}$ Overbeek 1951, Turner et al. 1993). Extensive studies have been reported for the metabolic pathways of trans-cinnamic acid in bacteria and eucaryotic cells (Floss 1979, Gross and Zenk 1969, Rosazza et al. 1993). The nutritional effects (Wolffram et al. 1994), and biological functions of trans-cinnamic acid on bacteria, fungi, and animal cancer cells have been studied (Bitsch et al 1984, Forti et al. 1996, Ekmekcioglu et al. 1998). cis-Cinnamic acid also plays an important role in the physiology of plant cells and fungi for the significant differences in the activities of bioactive enantiomers/diastereomers produced by or active against living microorganisms which are vitally important to pharmaceutical, flavor, and food industries (Hess et al. 1975, Aheldon 1993). However, due to the stability and purity of this unavailable cis-cinnamic acid, the mechanisms of cis-cinnamic acid on nutritional, toxicological, and metabolic pathway in eucaryotic cells are virtually unknown (Sun and Traxler 1998).

To date, the separation of cis-, and trans-cinnamic acids by ion-exchange chromatography method (based on pKa differences) was originally developed by Lindenfors in earily days (Lindenfors 1957). However, the elution of trans-cinnamic acid from that ion-exchange resin is a very slow process caused by strong attraction between the trans-, and cis-isomers of cinnamic acid. The strong partition effects between the trans-isomer of cinnamic acid and the packing resin also play a major factor for the poor separation of the racemic mixtures from the study by Lindenfors (1957). This is the likely reason that the cis-isomers of cinnamic acid obtained on that experiments were not stable (Lindenfors 1958). The cis-isomers of cinnamic acid obtained by that ion-exchange method probably contain a significant amount of trans-isomers. This may be the reason that the production of cis-cinnamic acid has not been commercialized so far.

In this paper, a simpler and much more precise method (based on difference in polarity) was present to separate cis-, and trans-cinnamic acids by using HPLC coupled with a semi-preparative C_{18} column. The obtained pure cis-isomers were fully characterized by their melting points, UV spectra, FT-IR spectra, Mass spectra, and coupling constants (J). Moreover, the stabilities of these cis-isomers were also examined.

Material and Methods

Chemicals

trans-Cinnamic acid was obtained from Aldrich Chemical Co. (Milwaukee, WI) with the purities of 99.5%. All other chemicals were Baker Reagent (Phillipsburg, NJ) or Fisher Certified ACS (Fair Lawn, NJ). All solvents were obtained from Fisher Scientific Co. (Agawam, MA) and are HPLC or GC grade.

Preparation of cis- and trans- cinnamic acid racemic mixture

To prepare cis- and trans-racemic mixture of cinnamic acid from the commercially available trans-compounds, 0.5 g of trans-cinnamic acid was dissolved in 10 ml acetone (GC grade) and placed in a quartz tube under a photochemical chamber reactor (the Southern New England Ultraviolet Co., Middletown, CT) and irradiated for 2 hr on a rotating plate surrounded by 16 ultraviolet lamps. Each lamp emits $1.65x10^{16}$/sec/cm^3 photons at 2537 A^o. After irradiation, the acetone was removed by sparging with N_2 until a dry residue was obtained. These residues were redissolved in 10 ml of methanol (HPLC grade) and filtered through a 0.45 μm membrane to remove any particulates and stored in brown bottles prior to HPLC separation.

Separation of the cis- and trans- cinnamic acid racemic mixture by HPLC

The Waters HPLC system (600E multisolvent delivery system) coupled with a 991 photodiode array detector (PDA) and a semi-preparative C_{18} reverse phase column (21.4 mm OD x 250 mm, Rainin instrument Co., Woburn, MA) were used to separate the cis- and trans-isomers. Methanol and deionized water mixtures were used as mobile phases for the separation of racemic mixture of cinnamic acid. In order to get a baseline separation, different mobile phase compositions and conditions were tested. The cis- and trans- isomers were detected by a Waters PDA detector set at 270 nm and scanned from 190-400 nm. The peak purity was monitored by a PC with the Millennium 2010 chromatography manager program. The eluents of the cis-isomers were collected in 500 ml round bottom flasks with a cover of aluminum foil. The mobile phase was removed by a rotatory evaporator under a 20 psi vaccum at 65°C.

Characterization and identification of the cis- and trans-cinnamic acid

The melting points of cis- or trans-cinnamic acids were determined by a MeltTemp (Laboratory Devices, Cambridge, MA). A 10 mg sample of each isomer was packed into a capillary tube and then transferred to the MeltTemp machine. The temperature was increased at a rate of 5°C/min until the crystals changed from solid to liquid phase. The temperature at this phase transition point was recorded as melting point (degree Celsius) for the sample.

The coupling constant (J) of the two protons (#2 and #3) on the unsaturated double bond which caused the difference between trans-, and cis-isomers were determined by one dimensional proton NMR (Nuclear Magnetic Resonance spectroscopy, Jeol Scientific Co., MA). The one dimensional proton NMR was set at $26^{\circ}C$ with a off set of 6 ppm. Tetramethylsilane (TMS) was used as the reference standard for the proton NMR. A 25 mg sample was dissolved in 1.5 ml of $CDCl_3$ in a condensed glass tube (Wilmad glass Co., Buena, NJ) and scanned to determine the total proton number, chemical shift (δ) and J. A HP 5980 GC coupled with a HP 5971 mass detector was used to identify the compounds. The GC column was a 30 m X 0.53 mm DB5 megabore column (J&W Scientific, Folsom, CA). The injection volume was 1 μl of 1 mg/ml of sample. The helium gas was used as carrier gas at a flow rate of 1 ml/min with a split ratio of 1:25. The temperature program was at an initial temperature $50^{\circ}C$ and programmed at $10^{\circ}C$/min to $300^{\circ}C$. The temperature for detector were set at $280^{\circ}C$. The mass detector was set at an Electrom Impart Ionization (EI)= 70 ev and scanned from 40 to 400 m/e (mass to charge ratio) at 1 scan/sec. One mg of sample was dissolved in 1 ml methylene chloride or methanol and saved for GC-Mass analysis. For all FT-IR determination, a 3 mg sample of a purified isomer was blended with 5 mg of crystalline KBr and pressed at 15,000 psi to form a plate. The plate was scanned from 400-4000 cm^{-1} by using a Perkin Elmer 1600 series FT-IR (Perkin Elemer Co, Newton, MA).

The stability test of the cis-cinnamic acid obtained by this method

The stability of the cis-isomer obtained after HPLC separation was monitored by the above HPLC system coupled with a Waters Nova-pak C_{18} reverse phase analytical column (3.9 mm ID X 150 mm) and a mobile phase of 52 methanol : 48 deionized water (containing 0.5% acetic acid) was used. The flow rate was 1 ml/min and the injection volume was 2 μl. The chromatographs were extracted at 270 nm for quantification. The change in peak area was used as an index for the stability of the cis-isomer at the 1st and 365th day after preparation.

Results:

Separation of racemic mixture of cinnamic acids by HPLC with a semi-preparative C_{18} column

After 2 hr irradiation for trans-cinnamic acid in the photochemical chamber reactor, a ratio of 55 and 45 was determined to be the trans- and cis-form respectively, by using the HPLC with a semi-preparative C_{18} column as shown in Figure 1. In order to get a base line separation of racemic mixtures of cinnamic acids, different mobile phase composition, flow rate, and injection volume were tested. The mobile phase with a mixture of methanol (50%) and deionized water mixture (50%) under a flow rate of 8.5 ml/min was proven to be the best condition for the separation of cis- and trans-cinnamic acids (Figure 1). After the trans- and cis-isomer was separated by the semi-preparative C_{18} column, the cis-isomer was collected and used for the further chemical characterization and identification.

Characterization and identification of cis-, trans-cinnamic acids and chlorinated cinnamic acids:
In the present study, the trans-cinnamic acid was shown to be white crystals and the melting point was determined to be $128\pm0.5\ ^{\circ}C$ (Table 1) which is $5^{\circ}C$ lower than that of Merck Index (The Merck Index, 1996). The cis-cinnamic acid obtained in this study was found to be a transparent oil-type liquid (Table 1) at room temperature which was different from the result (white crystal) of the experiment conducted by Lindenfor (1957). Based on the HPLC chromatogram as shown in Figure 2, the cis-isomer was eluted earlier than the trans-isomer and the retention time has 1.2 min difference. For the UV spectra, the different absorption maximum for cis-cinnamic acid (210.8 & 260.5 nm) and trans-cinnamic acid (215.7 & 275.5 nm) were similar to the results (cis-cinnamic acid: 210 & 265 ; trans-cinnamic acid: 217 & 275) from the experiment conducted by Lindenfors (1957). By using the Gas Chromatography, the cis-isomer (9.94 min) was also found to be eluted faster than the trans-cinnamic acid (10.95 min). Based on the results from the GC mass detector, the mass pattern for cis-cinnamic acid and trans-cinnamic acid were found to be similar, except two additional peaks (mass number 120 and 148) shown on the mass pattern of cis-cinnamic acid. According to the FT-IR spectra for cis-isomer and trans-isomer (Figure 3a and 3b), the major differences of the FT-IR spectra for cis- and trans-cinnamic acids are the peaks at the wavelength between 900 and 1,300 cm^{-1}. The results from the one dimensional ^1H NMR for the two isomers are shown on Table 1 expresses that the coupling constant (J) of the two protons (#2 and #3) for cis-isomer are 3.39 less than the trans-isomer.

Stability of the cis-cinnamic acid obtained from this study
The stability of the cis-isomers was determined by comparing the peak area of the cis-isomer on the HPLC chromatogram obtained at the first day with that of the cis-isomer at the 365th day. No significant change on the retention time and the peak area was found.

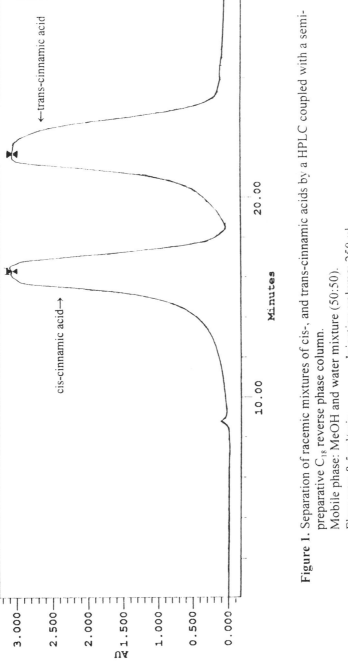

Figure 1. Separation of racemic mixtures of cis-, and trans-cinnamic acids by a HPLC coupled with a semi-preparative C$_{18}$ reverse phase column.
Mobile phase: MeOH and water mixture (50:50).
Flow rate: 8.5 ml/min.　　　Injection volume: 250 μl.

Table 1: Physical states, melting point of cis-, trans-cinnamic acids, and the chemical shift (δ), coupling constant (J) of these two isomers determined by one dimensional ^{1}H NMR.

Compounds / Determination	cis-cinnamic acid	trans-cinnamic acid
Physical states	transparent oil-type liquid	white crystal
Melting point	------	128±0.5 ^{o}C
$\delta(2)$ (ppm)	5.97	6.44
$\delta(3)$ (ppm)	7.05	7.98
J of carbon 2&3	12.44	16.03

$\delta(2)$: Chemical shift of the proton on the number 2 carbon on the unsaturated side-chain of cinnamic acid.

$\delta(3)$: Chemical shift of the proton on the number 3 carbon on the unsaturated side-chain of cinnamic acid.

J of carbon 2&3: Coupling constant of the proton between the number 2 and 3 carbon on the unsaturated side-chain of cinnamic acid.

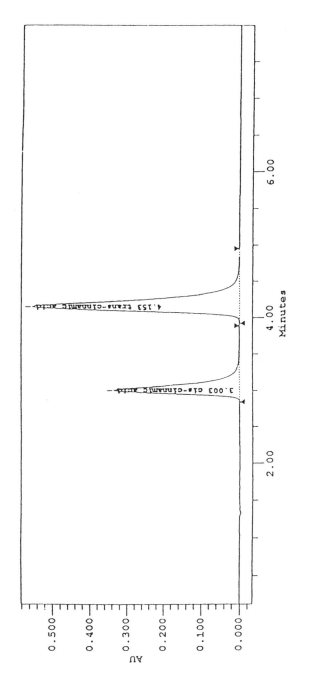

Figure 2. HPLC chromatogram of cis-, and trans-cinnamic acids.

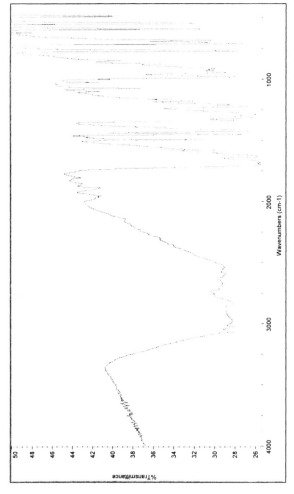

Figure 3a. FT-IR spectrum of trans-cinnamic acid

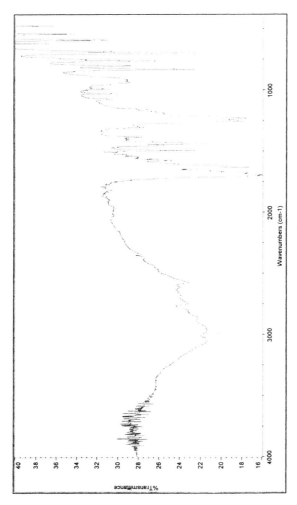

Figure 3b. FT-IR spectrum of cis-cinnamic acid

Table 1: Physical states, melting point of cis-, trans-cinnamic acids, and the chemical shift (δ), coupling constant (J) of these two isomers determined by one dimensional ¹H NMR.

Compounds / Determination	cis-cinnamic acid	trans-cinnamic acid
Physical states	transparent oil-type liquid	white crystal
Melting point	------	128±0.5 °C
δ(2) (ppm)	5.97	6.44
δ(3) (ppm)	7.05	7.98
J of carbon 2&3	12.44	16.03

δ(2): Chemical shift of the proton on the number 2 carbon on the unsaturated side-chain of cinnamic acid.

δ(3): Chemical shift of the proton on the number 3 carbon on the unsaturated side-chain of cinnamic acid.

J of carbon 2&3: Coupling constant of the proton between the number 2 and 3 carbon on the unsaturated side-chain of cinnamic acid.

Discussions

The only difference for the chemical structure between the cis- and trans-cinnamic acid is the protons on #2 and #3 carbon for cis-cinnamic acid are out of the plane at the double bond (Turner et al. 1993). This difference makes these two isomers totally different in their chemical, physical, and biological properties (Lindenfor 1957, Hess et al. 1975, Sun and Traxler 1998). Obvious differences in the pattern of mass spectra, FT-IR spectra, and ¹H NMR between cis-cinnamic acid and trans-cinnamic acid were found in the present study. The two additional peaks at the mass number of 120 and 148 on the mass spectrum of cis-cinnamic acid indicates that the cis-cinnamic acid is not easier to loss proton when compared with the trans-cinnamic acid. For the differences in the FT-IR spectrum between cis- and trans-cinnamic acid (Figure 3a and 3b), it may caused by the opposite position of the two protons on the double bond. The J of the two protons (#2 & #3) for cis-cinnamic acid is 3.39 less then the trans-cinnamic acid which is close to the result (3.3) calculated by Abraham et al. (1988). A smaller J for cis-cinnamic acid (Table 1) also indicates the distance for the two protons on the double bond of cis-cinnamic acid is closer than that of trans-cinnamic acid. In addition to the UV spectra and retention times that were used in early days to distinguish between these two isomers (Lindenfor 1957, 1958; Turner et al. 1993), the differences of their mass pattern, FT-IR spectra (Figure 3a, 3b), and J (Table 1) of cis-cinnamic acid and trans-cinnamic acid can be used as other important and better index to distinguish cis-isomer from the trans-isomer of cinnamic acids.

The results of the stability tests of the cis-cinnamic acid indicate that the conversion of cis-isomer back to trans-isomer do not happen suggesting the cis-isomers prepared by this new method are fairly stable in a brown bottle at room temperature for at least 365 day. Althrough, the results from the earlier experiments (Lindenfors 1957, 1958) showed that the cis-isomers of cinnamic acid and chlorinated cinnamic acids prepared by the ion-exchange method were not stable and the reason causing this unstability was uncertain. Based on the studies conducted by Lindenfor (1957, 1958), the elution buffer with 0.4 M acetic acid and 0.3 M sodium chloride in 75% (v/v) ethanol solution was used for the ion-exchange chromatograpy. The strong attraction force between the racemic mixtures and partition effects of the resins, the high acidity and chloride ion concentration were probably insufficient to result in a good separation for these racemic mixtures of cinnamic acids. The cis-cinnamic acid obtained by that ion-exchange chromatography method, therefore, may contain significant amount of trans-isomer, acetic acid and chloride ion. Moreover, the cis-cinnamic acid obtained from the ion-exchange chromatography method resulted in sodium salt (white crystal) which are different from the transparent oil obtained by the developed HPLC method in this study. These differences are the likely reasons that the cis-isomer of cinnamic acid obtained in earlier experiments were not stable. Whereas, the cis-cinnamic acid obtained by the HPLC method developed in the study was a pure acid and containing neither acetic acid nor chloride ion. By comparing with the ion exchange chromatography method developed by Lindenfors (1957 and 1958), the present HPLC method may represent a faster, simpler and much more precise method for the separation of the racemic mixture of cis- and trans-cinnamic acids.

Acknowledgements
This investigation was supported by the Rhode Island Agricultural Experiment Station project number 3858 (University of Rhode Island, USA).

References
Abraham, R. J.; Fisher, J.; Lofrus, P. Introduction to NMR spectroscopy; John Wiley & Sons: New York, 1988; pp 38-43.
Aheldon, R. Chirotechnology; Marcel Dekker: New York, 1993; pp 7-20.
Bitsch, A.; Trihbes, R.; Schultz, G. Compartmentation of phenylacetic acid and cinnamic acid synthesis in spinch. Physiol. Plant **1984**, 61, 617-621.
Chen, C-T; Sheu, S-J. Determination of glyrrhizin and cinnamic acid in commercial Chinese herbal preparation by Capillary Electrophoresis. Chin. Pharm. J. **1995**, 47-3, 213-219.
Ekmekcioglu, C.; Feyertag, J.; Marktl, W. Cinnamic acid inhibits proliferation and modulates brush border membrane enzyme activities in Caco-2 cells. Cancer Lett, **1998**, 128, 137-144.

Floss, H. G. The Shkimate pathway. In Biochemistry of plant phenolics; Swain, T., Harborne, J.B., Van Sumeze C. F., Eds.; Plenum Press: New York, 1979; pp 40-87.

Foti, M.; Piattelli, M.; Baratta, M. T.; Ruberto, G. Flavonoids, coumarins. and cinnamic acids as antioxidants in micellar system. structure-activity relationship. J. Agric. Food Chem. **1996**, 44, 497-501.

Gross, G. G.; Zenk, M. H. Reduktion aromatischer Sauren zu aldehyden und alkoholen in zellfrein system. 2. Reinigung und eigenschaften von aryl-alkohol:NADP-oxidoreduktase aus Neurospora crassa. Eur. J. Biochem. **1969**, 8, 420-425.

Hess, S. L.; Allen, P. J.; Nelson, D.; Lester, H. Mode of action of methyl cis-ferulate, the self-inhibitor of stem rust uredospore germination. Physiol. Plant Path. **1975**, 5, 107-112.

Lindenfors, S. Separation of cis-trans isomers by ion-exchange chromatography: 4chloro-cis-cinnamic acid and 2,4-dichloro-cis-cinnamic acid. Arkiv Kemi. **1957**, 10, 561-568.

Lindenfors, S. On the trans-cis isomerization of cinnamic acids by the action of ultraviolet rays: 3-chloro-cis-cinnamic acid and 3,5-dichloro-cis-cinnamic acid. Arkiv Kemi. **1958**, 12, 267-279.

Rosazza, J. P. N.; Huang, Z.; Dostal, L. Mechanism of ferulic acid conversions to vanillic acid and guaiacol by *Rhodotorula rubra*. J. Biol. Chem. **1993**, 268, 23954-23958.

Sun, F.M.; Traxler, R. W. Bioconversion of cis-, trans-cinnamic acids and chlorinated cinnamic acids by resting cell cultures of *Rhodotorula rubra* Y-1529. Ph.D. dissertation of Uni. of Rhode Island: Kingston, 1998; pp 20-46.

The Merck Index (12th edition). Merck & Co: New Jersey, 1996; pp 2363.

Turner, L. B.; Mueller-Harvery, I.; Mcallan, A. B. Light-induced isomerization and dimerization of cinnamic acid and derivatives in cell wall. Biochem. **1993**, 303, 791-796.

van Overbeek J.; Blondeau, R.; Horne, V. Trans-cinnamic acid as anti-auxin. Am. J. Botany **1951**, 38, 589-595.

Wolfrram, S.; Weber, T.; Grenacher, B.; Scharrer, E. A Na+-dependent mechanism is involved in mucosal uptake of cinnamic acid across the jeiunal brush border in rats. Nutr. Metab. **1994**, 113, 1300-1308.

Author Index

Subject Index

Vitamin B6
  bioavailability, 144
  bioavailability in foods, 145
  calculating requirements relative to
    protein intake, 144
  five major forms, 144–145
  See also Pyridoxal 5'-phosphate
    (PLP); Pyridoxal (PL)

W

Water, adiabatic compression, 15
Wheat, lysine, 144

Y

Yuba (soymilk film)
  changes in isoflavone content during
    processing, 82–83
  preparation, 75
  See also Soybean products

RETURN TO: **CHEMISTRY LIBRARY**
100 Hildebrand Hall • 510-642-3753

| LOAN PERIOD | 1 | 2 | **1-MONTH USE** |

1-MONTH U **ONTH USE**

ALL BOOKS **AFTER 7 DAYS.**
Renewals ma using GLADIS,
type **inv** foll

DUE A